Instructor's Manual for Hartmann's

Astronomy: The Cosmic Journey
Fourth Edition

by

John L. Safko
University of South Carolina

Wadsworth Publishing Company
Belmont, California
A Division of Wadsworth, Inc.

Printed in the United States of America   49

1 2 3 4 5 6 7 8 9 10——93 92 91 90 89

ISBN 0–534–09577–1

Preface

Astronomy has evolved rapidly in recent decades. Every few years, it seems, some major discovery has caused reassessment of one theory or another. A recent article on publishing pointed out that astronomy text publishers risk having obsolete books before finishing their first press run! For just this reason, the text takes the long view and states the uncertainties of some current mysteries and controversies rather than endorsing the latest interpretation of research of the last few years. For the same reasons, some of the problems in the text are worded so that there is not necessarily one correct answer; the intention is to get students thinking about the material. Each chapter of this manual has two sections: Answers to Problems in the Text and Sample Test Questions. The answers provided for some of these problems should be considered only as a guide to possible responses; instructors will have to decide for themselves what range of answers they will accept.

The Sample Test Questions are of four types: True-False, Multiple Choice, Essay, and Word Practice (or fill-in-the-blank). The questions attempt to cover most of the points raised in the text; again, instructors will have to select those questions most appropriate to the material emphasized in the classroom. Answers are provided for the True-False, Multiple Choice, and Word Practice questions, but Essay questions have only text page numbers as references for the answers. Most of the terms in the Word Practice are from the Concepts List at the ends of the text chapters. The few that are from elsewhere in the chapters are marked by an asterisk (*). Questions on material covered in the optional math sections are denoted by an asterisk.

Since we are trying to develop the most useful form for this supplement, we invite instructors' comments and criticisms and corrections for any errors in the answer sections.

William K. Hartmann   John L. Safko
Tucson, Arizona   *Columbia, S.C.*

Contents

Invitation to the Cosmic Journey

Answers to Problems in the Text

1. Each student will have his or her own answer to this question.

Sample Test Questions

True-False

1. The definition of astronomy today is the same as it always has been. F
2. This text treats scientific research as being like climbing a tree, starting near the trunk and proceeding to ever more specialized branches of knowledge. F
3. Our galaxy contains open star clusters. T
4. 3×10^7 is the same as 30,000,000. T
5. The unit of length in the SI system is the kilogram. F
6. We are on a cosmic journey, and astronomy is the process of finding out where we are. T
7. Our Sun is a star. T
8. A typical star cluster is bigger than the galaxy. F

Multiple Choice

1. Scientific research can be likened to
   - A. climbing up the tree of knowledge
   - *B. climbing down the tree of knowledge
   - C. riding a sled down a steep hill
   - D. climbing stairs
   - E. none of these

2. This text treats astronomy as
   - A. a narrow set of academic observations
   - *B. a voyage of exploration with practical effects on humanity
   - C. the science of casting horoscopes
   - D. both B and C
   - E. none of the above

3. The _____ is everything that exists.
   - A. nebula
   - B. Sun
   - C. galaxy
   - *D. universe
   - E. solar system

4. A _____ is a swarm of billions of stars.
   - A. universe
   - *B. galaxy
   - C. nebula
   - D. solar system
   - E. globular star cluster

1

5. A globular star cluster is
  A. everything that exists
  B. a swarm of billions of galaxies
 *C. a spheroidal mass of hundreds of thousands of stars
  D. a cloud of dust and gases

6. 5000 written in power-of-ten notation is
  A. 5×10^5 D. 5×10^{-3}
  B. 5×10^4 E. 5×10^{-4}
 *C. 5×10^3

7. The number 2.9×10^5 is the same as
 *A. 290,000   D. 0.000029
  B. 2,900,000   E. 0.0000029
  C. 290

8. Masses are expressed in kilograms in the _____ system.
  A. solar   D. natural
  B. English   E. star
 *C. SI

9. Lengths are expressed in _____ in the SI system.
  A. feet
 *B. meters
  C. centimeters
  D. kilometers
  E. inches

Essay

  1. Describe how scientific research is analogous to working our way down a tree. (See p. 2.)
  2. Give the text's definition of astronomy. (See p. 2.)
  3. Why does the author treat scientific research as analogous to working our way down a tree rather than up a tree? (See p. 2.)
  4. Arrange the following objects in order of increasing size: galaxy, Moon, Sun, Earth, star cluster, universe. (See p. 3.)

Word Practice

  1. A(n) _____ is a cloud of dust and gas.

  2. _____ is the process of finding out where we are.

  3. A(n) _____ is an enormous ball of gas, mostly hydrogen.

4. The largest grouping of stars is the _____. All of these groupings make the universe.

5. The _____ is everything that exists.

6. A grouping of stars found only in the spiral arms of galaxies is a(n)

   _____ _____ _____.

7. The _____ is the nearest natural body to the Earth.

8. The planet upon which we live is the _____.

9. A(n) _____ _____ _____ is a spheroidal mass of hundreds of thousands of stars.

10. The nearest star to us is the _____.

11. The _____ _____ is comprised mostly of the planets about our Sun.

12. The _____ notation is a convenient system for expressing large and small numbers.

13. The _____ _____ is the set of units that we will use to express physical dimensions.

Answers to Word Practice

1. nebula
2. astronomy
3. star
4. galaxy
5. universe
6. open star cluster
7. Moon

8. Earth
9. globular star cluster
10. Sun
11. solar system
12. power-of-ten
13. SI system

Chapter 1
Prehistoric Astronomy: Origins of Science and Superstition

Answers to Problems in the Text

1. a. Moves westerly (left) and comes closer to the horizon.
   b. Moves easterly (right) and rises higher above the horizon.
   c. Moves toward the horizon and toward the east (right).
   d. Moves away from the horizon and toward the east (right).

   All of these motions result from the stars' moving counterclockwise in circles about the NCP.

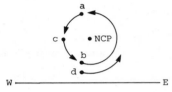

   Many students take the *just above* as a literal *at* and say "no motion."

2. a. Moves westerly (right) and comes closer to the horizon.
   b. Moves easterly (left) and rises higher above the horizon.
   c. Moves away from the horizon and toward the west (right).
   d. Moves away from the horizon and toward the east (left).

   All motions are now clockwise as seen by the observer, but east and west are now on different sides of the observer than they were in the Northern Hemisphere.

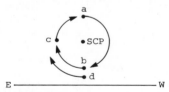

3. The elevation of Polaris is your latitude.
4. The circumpolar zone is the angular region about the celestial pole (NCP or SCP) of stars that do not set. At the equator the radius of the circumpolar zone is 0°, while at the North Pole it is 90°, so the radius, in general, is equal to your latitude. See Fig. 1-8.
5. Again, as in problem 4, at the equator you can see to the SCP, so you see 0°. At the north terrestrial pole you cannot see constellations closer than 90° from the SCP. The radius must be your latitude. See Fig. 1-8.
6. At any one point on the Earth, a solar eclipse will last no longer than 7.5 minutes, while a total lunar eclipse might last 1.75 hours. And the actual surface area of the Earth over which a total eclipse of the Moon can be seen is larger than the area over

4

which a total eclipse of the Sun can be seen. The surface area of the Earth on which you can see the Moon during a lunar eclipse is much larger than the area covered by the umbra of the Moon's shadow during a solar eclipse.

7. In general, no, for during an annular eclipse you are in the penumbral shadow. That is, you are observer C in Fig. 1-30. There is the rare possibility of an eclipse starting out total in one region of the Earth and then becoming annular at another point. At no time during the eclipse does the umbra touch that other point. When using Fig. 1-30, remember that it is not to scale. The umbral shadow is 162 times as long as the Moon's diameter.

8. If the Earth had no atmosphere, the Moon would completely disappear during a total lunar eclipse. During totality, light refracted by the Earth's atmosphere reaches the Moon to make a reddish cast. If, during what is a lunar eclipse as seen from Earth, we were on the Moon, there would be a total solar eclipse (as seen from the Moon). If the Earth had no atmosphere, it would appear as a black disk passing in front of the Sun. Because the Earth has an atmosphere, the true appearance is quite different. The atmosphere refracts (bends) light so the Earth appears to be surrounded by a brilliant ring of reddish light—red because that wavelength penetrates the Earth's atmosphere most easily. See Fig. 1-28.

9. It is probably easier to describe why either eclipse does not occur once each month. Because the Moon's orbit is tilted 5° to the Earth's orbit about the Sun, the shadow of the Moon usually falls above or below the Earth. Since the separation is 81 times the Earth's radius, 5° is enough for even the penumbra to miss the Earth. Only when the Moon is crossing the Earth's orbit when it is new (or full) can a solar (or lunar) eclipse occur.

10. a. Europeans became able to predict eclipses as early as 580 B.C., if not earlier. It would be fair to say that, although the Mayans did not become able to predict eclipses until A.D. 200–500, their knowledge was actually superior to that of the Europeans in common awareness.

   b. 600–800 years.

   c. This question can best be answered after Chapter 2. The Greeks understood the cause of eclipses many centuries before the Mayans. However, much of this knowledge was lost before the period under consideration. Both the Europeans and the Mayans were forced to use empirical tables that were corrected as they failed. The Mayan tables were better than the European tables.

   d. The Europeans were ahead in technology (for example, the American cultures did not have the wheel).

   e. No. Many other factors must be considered.

Advanced Problem

11. Referring to the diagram: Zenith–Celestial Equator angle equals the latitude of the observer. Thus, NCP–zenith angle equals 90° – latitude. Horizon–zenith angle is 90°, so λ (Horizon–NCP angle) is λ = 90° – (90° – latitude of observer). Hence, λ = latitude of observer.

Sample Test Questions

True-False

1. The NCP is the projection of the Earth's north axis of rotation. T
2. Polaris has always been the North Star. F
3. The ecliptic is the apparent path of the Moon across the sky. F
4. The zodiac is the region of the sky in which the planets are always found. T
5. A heliacal setting of a star is said to occur when a star sets soon after the Sun has set. T
6. The circumpolar zone is the region of the sky in which the stars set only after midnight. F
7. A solar eclipse can occur only when the Sun comes between the Earth and the Moon. F
8. If an eclipse occurs when the Moon's angular diameter appears smaller than the Sun's angular diameter, we have an annular eclipse at best. T
9. The umbra of the Moon is the region of the Moon's shadow in which none of the Sun can be seen. T
10. When the Moon is at the node of its orbit, it is north of the Earth's orbital plane. F
11. The line drawn between the two points where the Moon crosses the Earth's orbital plane is the line of nodes. T
12. The saros cycle is about 18 years, 11 days long. T

Multiple Choice

1. Some contemporary aborigines make calendar sticks to record
   A. historical events
   B. phases of the Moon
   C. binary star periods
   *D. both A and B
   E. all three, A, B, and C

2. The projections of the Earth's axis on the sky are the
   A. celestial equators
   B. zenith
   *C. celestial poles
   D. ecliptic

3. The brightest star nearest the Earth's celestial pole is currently
   *A. Polaris
   B. Vega
   C. Sol
   D. Ecliptus
   E. none of these

4. The path of the Sun among the stars is called the
   A. celestial equator
   B zodiac
   C. heliacal path
   *D. ecliptic
   E. none of these

6

5. The heliacal rising of a star occurs
   A. on the last day of each year when a star can be seen just before dawn
   *B. on the first day of each year when a star can be seen just before dawn
   C. on the last day when a star can be seen at dusk
   D. on the first day when a star can be seen at dusk
   E. on the day it is eclipsed by the moon

6. Many of our familiar constellations probably came from the
   A. Romans   D. Chinese
   B. Greeks   *E. Minoans
   C. Mayans

7. The precession of the Earth's axis is caused by
   A. lunar forces
   B. solar forces
   C. the attractions of the star Polaris
   *D. both A and B
   E. none of these

8. The precessional period of the Earth's axis takes about
   A. 13 years   *D. 26,000 years
   B. 37.5 days   E. 3×10^7 years
   C. 2600 years

9. The Sun crosses the celestial equator going north on or about March 21. This is
   known as the spring
   *A. equinox   D. zodiac
   B. solstice   E. crossing
   C. Stonehenge

10. Ancient buildings oriented east-west are said to have _____ orientation.
    A. a solstitial   D. a precessional
    *B. an equinoctial   E. a heliacal
    C. an ecliptical

11. Stonehenge is a(n) _____ oriented structure.
    *A. solstitially   D. precessionally
    B. equinoctially   E. heliacally
    C. ecliptically

12. If the Moon completely covers the Sun as seen by an earthbound observer, we have
    a _____ eclipse.
    A. total lunar   D. partial solar
    *B. total solar   E. both A and B
    C. partial lunar

13. For a lunar or solar eclipse to occur, we must have
    A. the line of nodes of the Moon's orbit pointing at the Sun
    B. the Moon at or near one node
    C. the Sun between the Earth and the Moon
    *D. both A and B
    E. all three, A, B, and C

14. The eclipse year is about 346.6 days rather than 365 days because
    *A. the nodes of the Moon's orbit precess
    B. the Moon's orbit is not perfectly circular
    C. misleading: The eclipse year is 365.24 days, as is the regular year.
    D. the lunar month is about 29 days
    E. both B and C

15. Where behind the Moon would you be able to see the following? __C__ .

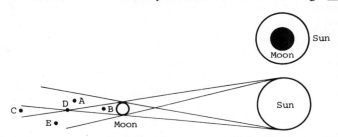

16. The saros cycle is completed when which two cycles repeat together again?
    A. the lunar period of 29.5 days and the year of 365.24 days
    *B. the lunar period of 29.5 days and the eclipse year of 346.6 days
    C. the lunar period of 29.5 days and the precessional period of 5000 years
    D. the menstrual period and the average month in the year
    E. none of these

17. The Mayan Sacred Round of 260 days was chosen so that _____ eclipses
    could occur in two rounds.
    A. one   D. four
    B. two   E. five
    *C. three

18. Even though solar eclipses occur more often than lunar eclipses, why is it more
    likely that you will see a lunar eclipse?
    A. It is easier to see eclipses at night.
    B. A lunar eclipse is visible for only a small region of the Earth.
    *C. A solar eclipse is visible for only a small region of the Earth.
    D. misleading: Lunar eclipses occur more often.
    E. Both B and D.

8

Essay

1. Describe some of the probable motivations of early astronomy. (See p. 9.)
2. Briefly discuss the probable origin of the constellations according to Ovenden's reasoning. (See pp. 14–15.)
3. Discuss the difference between equinoctial and solstitial orientation of buildings. (See pp. 21–25.)
4. Describe the basic structure of Stonehenge and its probable use. (See pp.19–22.)
5. Compare the structure and possible use of
   a. Temple of Amon-Re, Karnak, Egypt
   b. Casa Grande, Arizona (See pp. 19–20.)
6. Distinguish between astrology and astronomy. (See pp. 25–28.)
7. Describe at least two tests that astrology fails. (See pp. 27–28.)
8. What is the saros cycle? (See pp. 33–34.)
9. Make a sketch of the Sun and Moon and show the shadow regions behind the Moon. Label the regions in which more of the Moon can be seen as (a), label an annular Sun as (b), and label a partial eclipse as (c). (See p. 34.)

Word Practice

1. The slow rotation of the plane of the celestial equator is called _____.

2. Stars that never set throughout the year are said to lie in the _____ _____.

3. The _____ _____ of a star occurs on the first day of each year when the star can be seen just before dawn.

4. The region of the sky that contains stars that rise and set is termed the _____ _____.

5. Globes that circle the Sun, such as our own Earth, are called _____.

6. An eclipse in which the Sun can be seen all around the edges of the Moon is termed a(n) _____ _____ _____.

7. Groupings of stars used as a memory aid are called _____.

8. The _____ are points where the Moon's orbit crosses the ecliptic.

9. The current North Star (or pole star) is _____.

10. The point directly overhead is the _____.

11. The region of the Moon's shadow where the Sun cannot be seen is the _____.

12. The line of intersection of the Moon's (or any planet's) orbit with the ecliptic plane is called the _____ _____ _____.

9

13. The plane of the Earth's orbit is the _____. This is the apparent path of the Sun against the stars.

14. The tilt of the Earth's rotation axis to the plane of the ecliptic is the

_____ _____ _____ _____.

15. The star nearest to the north celestial pole is the _____ _____.

16. When the Sun or Moon disappears while above the horizon, we have a(n)

_____.

17. The projection of the Earth's axis of rotation onto the sky produces the

_____ _____.

18. When the Earth's shadow falls on the Moon, we have a(n) _____

_____.

19. When the Sun is completely hidden by the Moon, we have a(n) _____

_____ _____. Thus, we are in the Moon's umbra.

20. The zone 9° on both sides of the ecliptic (18° total width) is called the

_____.

21. _____ occur when the Sun crosses the celestial equator going north or south.

22. A(n) _____ _____ _____ occurs when the Moon hides only part of the Sun.

23. When the Moon passes in front of the Sun, we have a(n) _____ eclipse.

24. A(n) _____ _____ occurs on the last day when a star can be seen at dusk.

25. The projection of the Earth's equator on the sky is the _____

_____.

26. The _____ is the region of the Earth's or Moon's shadow where the Sun is only partially covered. See D and E of Fig. 1-28.

27. An imaginary line passing through the north celestial pole and your zenith is

the _____.

28. When the Sun is farthest north or south of the celestial equator, we have a(n)

_____.

29. The hour circle through the zenith is the local _____.

30. The local meridian passes through the _____ and the celestial poles.

10

31. _____ causes the vernal equinox to appear to move along the celestial equator.

32. A(n) _____ occurs when the Sun is on the celestial equator.

Answers to Word Practice

 1. precession
 2. circumpolar zone
 3. heliacal rising
 4. equatorial zone
 5. planets
 6. annular solar eclipse
 7. constellations
 8. nodes
 9. Polaris
10. zenith
11. umbra
12. line of nodes
13. ecliptic
14. cause of the seasons
15. North Star
16. eclipse

17. celestial poles
18. lunar eclipse
19. total solar eclipse
20. zodiac
21. equinoxes
22. partial solar eclipse
23. solar
24. heliacal setting
25. celestial equator
26. penumbra
27. meridian
28. solstice
29. meridian
30. zenith
31. precession
32. equinox

Chapter 2
Historic Advances: Worlds in the Sky

Answers to Problems in the Text

1. Not until the days of space exploration could a theory that placed the Earth at the center of the universe be completely ruled out. Until then, it was always possible that the laws controlling the planets and stars were different from those controlling the Earth, although great simplification and unification were to be gained by assuming these laws were the same. Thus, whereas now that we have sent out spacecraft we cannot support by any scientific means the Earth-centered universe, the early theorists had the opposite problem: There was no direct evidence to show that the Earth moved.

   a. There was no observational basis for not putting the Earth at the center. In fact, theorists could argue that because they felt no motion, the Earth was obviously at rest. Those who asserted that the Earth was in motion were probably in the "lunatic fringe" of their day because they lacked any direct observational support for their ideas.

   b. Either theory could fit the available data. The lack of parallax of the stars when observed by the naked eye supports the Earth-centered universe. Only the invention of the telescope enabled astronomers to see the small motions (parallax) of the stars caused by the Earth's revolving about the Sun.

   c. The Greeks did make good arguments to show that the Moon revolves about the Earth and that the planets and Sun do so as well. But they could neither *prove* nor *disprove* their theories.

2. Remember that the terminator is not the shadow of the Earth. The Sun lights half the Moon at all times. The terminator is the division between the lit and the unlit halves. We see a terminator only because of the angle at which we view the Moon.

   a. The terminator is usually seen as curved because we are not looking straight down on it. That is, if the terminator were imagined as a plane bisecting the Moon, our viewpoint of the Moon is usually at any angle to that plane.

   b. The terminator is a straight line at half phase, that is, when the Sun and Moon form a right angle (90°) when seen from the Earth.

   c. The terminator is not seen when the Moon is new because nothing is seen then (assuming no eclipse). When the Moon is full, the terminator is the edge of the disk of the Moon. During an eclipse, the edge of the occulting disk is the terminator.

3. The Greeks could measure the cyclic motions of the planets, Moon, and Sun using sighting devices similar to the ones described in Chapter 1. They could also rank the stars by apparent relative brightness using only the eye, but they had no information on the relative distances or physical nature of the stars. All observations were limited to visible light. Modern astronomers can use wavelengths from gamma rays up to radio waves. Telescopes are available to collect light so we can see fainter objects and magnify images. We can even reduce light to its various wavelengths. Modern techniques can more accurately measure the relative brightness, distances, sizes, compositions, and motions. The space program allows us to actually visit some of the nearer astronomical objects.

4. Most students will say yes to these questions. Examples can range from such dramatic events as eclipses, meteorite falls, and comets to the less dramatic but more regular seasonal change in the heavens. Astronomical observations suggest that humanity is subject to cosmic forces—forces that affect human behavior and events.

5. Aristotle's belief helped in the sense that raw observations must be structured to be interpreted. (Indeed, a structure is needed even before observations are made—how else to decide what is worth observing?) Aristotle's faith in symmetry and "perfection" provided such a structure. But his faith hindered him in that it caused him to ignore or explain away observations that did not support his structure. (This is always a danger when the faith is too detailed.) Thus, he placed the Earth at the center of the universe.

6. Looking at Fig. 2-7, which shows the Earth and the Moon, we see that the farther away we are from the object casting the shadow, the smaller the shadow is. This is true whenever the object providing the light (the Sun) is physically larger than the object casting the shadow (the Moon). It does not matter how far apart the objects are. In brief, the answer is because the Sun is bigger than the Moon.

7. To have discovered that the north celestial pole (or the stars near the vernal equinox) had moved, he needed records at least 500 to 1000 years old because all observations were by the naked eye.

8. The answer will vary from student to student, but the most common answer should be significant. Because most ancient knowledge was lost, astronomers were forced to start over again. Whether this was good or bad is another question. Some students will argue that such events are insignificant because they only delayed but did not stop the development of astronomy.

9. Remember, the zenith is the point overhead. At the equator, the zenith lies on the celestial equator. On the equator, the Sun passes through the zenith only in late March and September (at the equinoxes).

Advanced Problems

10. The Sun is at most 23°5 north of the celestial equator. Let λ be your latitude. Then the NCP lies λ up from the north, while the celestial equator lies $90° - \lambda$ up from the south $[180° - (90° + \lambda)]$. So the Sun is at most $90° - \lambda + 23°5$ up from the south. If $\lambda = 40°$, the Sun rises at most $73°5$ from the south—the answer is no. If $\lambda = 23°5$, then the Sun can be at the zenith, and it is so only on the summer solstice (about June 21).

11. a. Use the small angle equation. α must be in seconds of arc so $1/2° = 30' = 1800"$, while $D = 150$ million km $= 1.5 \times 10^{11}$ m. $\alpha"/206,265 = d/D$ becomes $d = (\alpha" \times D)/206,265$. $d = (1.8 \times 10^3 \times 1.5 \times 10^{11}$ m$)/206,265$. $d = 1.31 \times 10^9$ m $= 1.31 \times 10^6$ km.

    b. Yes, he could measure α, so $d/D = \alpha"/206,265$.

    c. They had no technique to separately measure either d or D. To get D independently, they would need to measure an angle about 8", whereas without telescopes they could at best measure to a few minutes of arc. The latter comment is extra information probably not known by the students at this point.

13

12. a. The eye can resolve 2', so if there are 435 pixels in a 40° angle, then each pixel is 40/435 = .092, which is about 5'.5. We can approximate how close this would be for the human eye by using the small angle equation, although 40° is clearly not a small angle. Converting 40° to seconds of arc and assuming a 21-inch TV, we get 3 inches. At 3 inches we can clearly see pixels. At most viewing distances, the pixels will not be resolved.

   b. European TVs will look better at close distances, while British TVs will look worse. At usual viewing distances, the choice will not make much difference. If the TV is used for a computer readout or other close work, the European system will be superior.

13. Again use the small angle equation. You are given $\alpha'' = 1''$ and $D = 400,000$ km = 4×10^8 m, so $d = (1'' \times 4 \times 10^8$ m$)/206,265 = 1.9 \times 10^3$ m = 1.9 km.

14. Remember, on Earth the angle of the NCP is your latitude. If on the satellite 1 km corresponds to a star movement of 1° (and hence the local NCP moves 1°), then 360 km would be a 360° motion. Thus, the circumference would be 360 km. The diameter is the circumference divided by π (3.14), so the diameter is about 115 km.

15. The "horns" must point away from the Sun. Since we can see the horizon, the Sun must be up, and since the crescent is narrow, we are not too far from new Moon so the Sun is not post noon. Conclusion: In the morning when the Moon is past third quarter.

Sample Test Questions

True-False

1. Cosmologies are theories of the origin and nature of the universe. T
2. The Egyptians did not believe there was a "divine order" to the universe. F
3. Early Greek astronomy began around 600 B.C. T
4. Thales of Miletus was an early astronomer of the Greek world. T
5. Anaximander is known for his eclipse predictions. F
6. The lunar terminator separates the lit side of the Moon from the unlit side. T
7. Anaxagoras deduced the true cause of eclipses. T
8. Aristotle believed the Moon was farther away than the Sun. F
9. Aristarchus deduced the approximate relative sizes and distances of the Sun and Moon. T
10. Aristotle deduced the relative sizes of the Sun and Moon. F
11. Eratosthenes discovered the precession of the equinoxes. F
*12. The number of digits known for certain in a quantity is the number of significant figures of that quantity. T
13. The epicycle theory is especially associated with Ptolemy. T
14. Al-Battani was an early Indian astronomer. F
15. Chinese astronomy began much later than Greek astronomy. F

Multiple Choice

1. The early Greek astronomer who is first recorded as believing that the heavenly bodies are spherical in shape is
   A. Brahe
   B. Philolaus
   C. Aristotle
   *D. Pythagoras
   E. Ptolemy

2. An early Greek who gave a clear and correct explanation of the phases of the Moon was
   A. Ptolemy
   B. Copernicus
   C. Brahe
   D. Kepler
   *E. Aristotle

3. The Ptolemaic cosmology of deferents and epicycles was based on the ideas of Hipparchus.
   *A. correct
   B. wrong: Ptolemy thought it up all by himself.
   C. wrong: It was based on the work of Kepler.
   D. wrong: It was based on the work of Copernicus.
   E. wrong: It was based on the work of Newton.

4. The cosmological scheme in which the Earth is at the center of the universe and the planets, Sun, and Moon are on circles whose centers are moving in circles about the Earth was devised by the Greek astronomer
   A. Copernicus
   B. Aristotle
   C. Hartmann
   D. Philolaus
   *E. Ptolemy

5. In Ptolemy's cosmology, a planet moves on the epicycle and the center of the epicycle moves around the Earth on the deferent circle.
   *A. correct
   B. wrong: The planet stays on the deferent circle.
   C. wrong: The Earth is at the center of the epicycle.
   D. wrong: Ptolemy put the Sun at the center.
   E. wrong: Both B and C are correct.

6. One of the first known Greek astronomers was
   A. Brahe
   *B. Thales of Miletus
   C. Aristotle
   D. Pythagoras
   E. Ptolemy

7. Aristarchus of Samos is famous for
   A. his discussion of the Moon's phases
   *B. his calculations of the relative distances of the Sun and Moon
   C. his catalog of nebular objects
   D. finding that the orbits of the planets were ellipses
   E. none of the above

8. Eratosthenes was the first astronomer to correctly measure the Earth's size.
  *A. correct
  B. wrong: Eudoxus
  C. wrong: Hipparchus
  D. wrong: Bernard
  E. none of the above

9. A theory of the origin and nature of the universe is termed
  *A. a cosmology
  B. a constellation
  C. an eclipse
  D. a deferent
  E. an epicycle

10. Thales of Miletus is best known for
  A. discussing a planet
  B. explaining the phases of the Moon
  *C. predicting an eclipse
  D. measuring the size of the Moon
  E. inventing the telescope

11. The line separating the lit side from the unlit side of the Moon is termed the lunar
  A. period
  B. phase
  C. epicycle
  *D. terminator
  E. node

12. The shift in the apparent position of a star due to the Earth's motion about the Sun is termed the star's
  A. terminator
  *B. parallax
  C. deferent
  D. phase
  E. none of these

13. Precession was discovered by
  A. Eratosthenes
  *B. Hipparchus
  C. Ptolemy
  D. Thales of Miletus
  E. none of these

14. The *Almagest* was written by
  A. Eratosthenes
  B. Hipparchus
  *C. Ptolemy
  D. Thales of Miletus
  E. none of these

15. The Arab astronomer best known in medieval Europe was
  A. Muhammad Almagest
  *B. Muhammad Al-Battani
  C Eratosthenes
  D. Aristotle
  E. none of these

16. The main center of Indian astronomy was at
  A. Almagest
  B. Al-Battani
  *C. Benares
  D. Alexandria
  E. Samos

16

17. Chinese astronomers believed
    A. the Earth was at rest
 *B. the Earth was in motion
    C. the stars were most important for sailing

*18. If a building 500 meters away has an angular height of 1°, approximately how tall is it?
    A. 5 m   D. 9 m
 *B. 9 m   E. 500 m
    C. 12 m

*19. When Mars is closest to the Earth, its angular diameter is about 85" and its distance is about 7.7×10^7 km. The approximate diameter of Mars is
    A. 6×10^3 km   *D. 3.2×10^3 km
    B. 300 m   E. 5×10^3 km
    C. 77×10^3 km

*20. (easy numbers) A space satellite 1 meter in diameter is circling the Earth. It is observed to have an angular size of 2.06 seconds of arc. The distance to the satellite is
    A. 10^5 cm   *D. 10^7 cm
    B. 10^6 cm   E. 2.06×10^3 cm
    C. $(1/2.06) \times 10^3$ cm

*21. (easy numbers) A 2-meter-tall man is observed to have an angular height of 0.57° = 2060". How far away is he?
 *A. 200 m   D. 570 cm
    B. 200 km   E. 5.7×10^3 m
    C. 200 cm

*22. The difference in direction between two points as measured from a specified third point is the
    A. parallax
    B. linear measure
 *C. angular measure
    D. constant of proportionality
    E. separator

*23. If α is the angular size of an object in seconds of arc, d its real size, and D the distance to the object, these quantities are related by
    A. $\alpha d = D/206{,}265$ D. $\alpha/206{,}265 = dD$
 *B. $\alpha/206{,}265 = d/D$ E. none of these
    C. $\alpha/206{,}265 = D/d$

17

Essay

1. Discuss the effects that early East Indian, Chinese, Polynesian, and American Indian astronomy had on European astronomy. (See pp. 54–57.)
2. Name and discuss the contributions of three early Greek astronomers. (See pp. 44–52.)
3. Name and discuss the contributions of two late Greek (Alexandrian) astronomers. (See pp. 52–53.)
4. What is resolution? (See p. 43.)
5. Discuss the units of angular measurement. (See p. 42.)
6. What is an angle? (See pp. 46–47.)

Word Practice

1. A shift in apparent position of an object due to your motion is termed

   _____.

2. Over the course of the year, stars near the ecliptic appear to move in a straight line. This is the _____ _____.

3. 1/60 of a minute of arc is a(n) _____ _____ _____.

4. _____ _____ is a measure of the difference in direction between two points as seen from a specified third point.

*5. _____ was a Greek astronomer who wrote the *Almagest*.

*6. One of the first known Greek astronomers was _____ _____ _____. He predicted the eclipse discussed in Chapter 1.

7. Theories about the origin and nature of the universe are called _____.

8. _____ _____ is a measurement of the distance between two points.

9. The _____ _____ was a famous library in Egypt.

10. The _____ _____ _____ is a drift in the location of the celestial poles with respect to the stars, caused by forces produced by the Sun and Moon.

11. The _____ _____ _____ relates an object's angular size, linear size, and distance.

*12. _____ was a Greek who is credited with deducing the true cause of eclipses.

*13. The _____ is a circle whose center moves in a circular orbit centered near the Earth (Ptolemaic cosmology).

*14. _____ was a Greek who proposed that the matter from which things were made was composed of an eternal substance.

15. The _____ is a measure of the finest detail that can be seen in an image.

16. The _____ is the line separating the light and dark sides of the Moon (or any planet).

17. _____ is a process of learning about nature by formulating questions and answering them through observation and experiment.

*18. The _____ is a circle upon which the epicycle moves (Ptolemaic cosmology).

19. Sixty seconds of arc make up one _____ _____ _____.

*20. A famous Arab astronomer was _____.

*21. _____ was an astronomer who proposed that the Sun was farther from the Earth than the Moon and who explained the phases of the Moon.

*22. The digits known for certain in a quantity are the _____ _____.

23. 1/360 of a circle is a(n) _____.

24. Ptolemy proposed the _____ _____ to predict positions of the Sun, Moon, and planets.

*25. _____ was a Greek astronomer who proposed a spherical shape for heavenly bodies.

*26. The name of _____ is associated with determining the size of the Earth.

*27. _____was a Greek famous for his star catalog and observations.

28. The first astronomer to propose the heliocentric hypothesis was _____.

1. parallax
2. parallactic shift
3. second of arc
4. angular measure
5. Ptolemy
6. Thales of Miletus
7. cosmologies
8. linear measure
9. Alexandrian library
10. precession of equinoxes
11. small angle equation
12. Anaxagoras
13. epicycle
14. Anaximander
15. resolution
16. terminator
17. science
18. deferent
19. minute of arc
20. Al-Battani
21. Aristotle
22. significant figures
23. degree
24. epicycle theory
25. Pythagoras
26. Eratosthenes
27. Hipparchus
28. Aristarchus

Chapter 3
Discovering the Layout of the Solar System

Answers to Problems in the Text

1. Counterclockwise; clockwise.
2. Mercury; Pluto.
3. No.

4. Venus can be full phase only when it is on the far side of the Sun from the Earth.
5. Earth.
6. No. The effect is purely one of appearance and occurs when the observation point (Earth) is overtaking Mars.
7. Not clear; presumably we are only asking them to be nearly on a straight line—their being exactly on a straight line is extremely rare. If they are nearly on a straight line, we see Venus at sunset on the western horizon (inferior conjunction) and Mars and Jupiter rising in the east nearly together (in conjunction with each other while at opposition). If all are exactly on a straight line, we see Venus pass in front of the Sun (transit) and Mars pass in front of Jupiter (occult Jupiter). If we are on Mars, we see Earth and Venus on the western horizon at sunset (in inferior conjunction) and Jupiter rising on the eastern horizon (in opposition), and if the line is exactly straight, then from Mars we see Earth and Venus transit the Sun. From Jupiter, we see Venus, Earth, and Mars on the western horizon at sunset (in inferior conjunction); if they are on an exact straight line, we see all three transit together as one.
8. Any answer may be acceptable provided it is properly justified. Provided below are several viewpoints and their justifications for the questions involved.

Revolutionaries and Why
a. Perhaps not: They were only following the evolving scientific knowledge, although they were aware of the strong pressures against their findings.
b. Definitely: Copernicus would not publish until he was on his deathbed. Galileo recognized that he was undermining the power structure. Both, by their actions, showed they recognized the revolutionary nature of their ideas.

Scientific vs. Political: Similarities and Differences
a. Similarities: There is an established power structure, new ideas about how things should be done, and newcomers who force their ideas to attention. Both scientific and political revolutions have gestation periods of about a generation in length. Many scientific revolutions have succeeded because those in the establishment died of old age, not because they became convinced by the "facts."

b. Differences: A political revolution much more strongly involves opinion, social style, and other factors that do not involve verifiable fact. Scientific revolutions are based on verifiable observations (independent of style or opinion).

Role of Factual Discoveries, Strong Personalities, and so on
Probably all agree that in the long run a scientific revolution is much more dependent upon factual discoveries than is a political revolution. Both revolutions are strongly affected by strong personalities and publicity. A strong personality, by promoting or rejecting new data, may help or hinder a scientific revolution. Koestler, in his book *The Sleepwalker,* argues that Galileo's strong personality actually hardened opposition to the Copernican theory.
a. The long run referred to above may be very long—witness the tenacity of Soviet biology against overwhelming disconfirming "facts." Strong personality and publicity were the weapons of resistance to change.
b. But it must be stressed that in the long run the verifiable facts win out in a scientific revolution, whereas in a political revolution there may, indeed, be no verifiable facts.

Advanced Problems

9. $\dfrac{\alpha"_{\delta}}{2.06 \times 10^5} = \dfrac{6.79 \times 10^6 \text{ m}}{6 \times 10^{10} \text{ m}} = 1.13 \times 10^{-4}; \ \alpha"_{\delta} = 23"$

$\dfrac{\alpha"_{\mathcal{4}}}{2.06 \times 10^5} = \dfrac{1.43 \times 10^8 \text{ m}}{6.3 \times 10^{11} \text{ m}} = 2.27 \times 10^{-4}; \ \alpha"_{\mathcal{4}} = 47"$

That is, Jupiter is about 30 times bigger than Mars and 10 times farther away. Jupiter appears larger.

10. We are given that Mars has a diameter of 6787 km = 6.8×10^3 km and that it approaches within 60 million km = 6.0×10^7 km. We are asked the smallest surface feature we can resolve on Mars if our resolution is 1/2". Let $D = 6.0 \times 10^7$ km be the distance to Mars and d be the size of the smallest feature we can resolve. The small angle equation gives

$\dfrac{\alpha}{206,265} = d/D$ or $1/2(2.1 \times 10^5) = d \, (6.0 \times 10^7 \text{ km})$

$d = 1.4 \times 10^2$ km = 140 km.

Sample Test Questions

True-False

1. Jupiter is closer to the Sun than Saturn is. T
2. Neptune is closer to the Sun than Uranus is. F
3. Superior planets are closer to the Sun than the Earth is. F
4. Only superior planets can be in opposition. T
5. A planet in opposition rises at sunset. T
6. The Ptolemaic model had the Sun at the center of the universe. F
7. One way of stating Occam's razor is, "The best explanation is the simplest." T

8. The Copernican theory put the Sun at the center of the solar system. T
9. Copernicus wrote *De Revolutionibus.* T
10. Tycho Brahe was a Czech astronomer. F
11. Tycho Brahe is best known for his observations. T
12. Kepler first studied the planet Mars to develop his laws. T
13. Galileo is the first known astronomer to use the telescope. T
14. Newton deduced the laws of planetary motion from more basic laws of general motion. T
15. Kepler found his laws of motion empirically rather than by deducing them. T
16. Newton's third law may be stated as follows: The force on a body is equal to the product of its mass times its acceleration. F

Multiple Choice

1. The superior planets are, in order of increasing average distance from the Sun:
    A. Mercury, Venus, Mars, Jupiter
    B. Mars, Jupiter, Saturn, Uranus, Pluto, Neptune
  *C. Mars, Jupiter, Saturn, Uranus, Neptune, Pluto
    D. Mars, Jupiter, Saturn, Neptune, Uranus, Pluto
    E. Saturn, Neptune, Jupiter, Pluto, Uranus, Mars

2. Which of the planets is closer to the Sun than the Earth is?
    A. Mars   D. Saturn
  *B. Venus   E. both A and B
    C. Jupiter

3. Which answer defines the term or concept *inferior planet?*
    A. actual period of revolution about the Sun with respect to the fixed stars
    B. has an orbit about the Sun larger than the Earth's orbit
    C. consists of observations and predictions
    D. apparent period of revolution with respect to the Sun
  *E. has an orbit about the Sun smaller than the Earth's orbit

4. The retrograde motion of Mars can best be explained as
    A. Mars reversing its motion
  *B. the Earth overtaking and passing Mars
    C. the Earth reversing its rotation
    D. the stars reversing their motion
    E. none of these

5. One theory that places the Earth at the center of the solar system with the Sun and planets revolving about it is the _____ model.
    A. Copernican   D. Brahe
  *B. epicycle   E. centered
    C. Occam

6. The *Almagest* was written by
   A. Kepler
   B. Copernicus
   C. Brahe
   D. Occam
   *E. Ptolemy

7. Match a description with the astronomer: Tycho Brahe
   A. made the first accurate measurements of the Earth's size using a well near Aswan and observations at Alexandria in Egypt.
   B. was the first astronomer to use a telescope to observe the phases of Venus.
   *C. (1546–1601) was famous for his accurate observations of positions of planets and stars; did not use a telescope.
   D. (1571–1630) was the first to discover that the orbits of the planets were ellipses.
   E. (1643–1727) was the first to successfully find physical laws that explain both terrestrial and celestial motion.

8. Match a description with the astronomer: Johannes Kepler
   A. made the first accurate measurements of the Earth's size using a well near Aswan and observations at Alexandria in Egypt.
   B. was the first astronomer to use a telescope to observe the phases of Venus.
   C. (1546–1601) was famous for his accurate observations of positions of planets and stars; did not use a telescope.
   *D. (1571–1630) was the first to discover that the orbits of the planets were ellipses.
   E. (1643–1727) was the first to successfully find physical laws that explain both terrestrial and celestial motion.

9. Match a description with the astronomer: Isaac Newton
   A. made the first accurate measurements of the Earth's size using a well near Aswan and observations at Alexandria in Egypt.
   B. was the first astronomer to use a telescope to observe the phases of Venus.
   C. (1546–1601) was famous for his accurate observations of positions of planets and stars; did not use a telescope.
   D. (1571–1630) was the first to discover that the orbits of the planets were ellipses.
   *E. (1643–1727) was the first to successfully find physical laws that explain both terrestrial and celestial motion.

10. The book *De Revolutionibus* was written by
    A. Ptolemy
    B. Kepler
    C. Newton
    D. Galileo
    *E. Copernicus

11. The deferent is
    *A. one of the parts of the epicycle model
    B. a bright comet
    C. the distance between the Earth and Sun
    D. the passing of a planet between the Earth and the Sun
    E. none of these

12. Kepler's second law says, essentially, that
    A. force equals mass times acceleration
    B. the square of the period is proportional to the cube of the semimajor axis
    C. for every action there is an equal and opposite reaction
    D. the orbits of the planets are ellipses with the Sun at one focus
    *E. the line from the Sun to a planet sweeps equal areas in equal times

13. The constant in Kepler's third law can be set to unity, provided we measure
    A. period in years   *D. both A and C
    B. period in days   E. both B and C
    C. distance in AUs

14. Kepler's approach to the problem of planetary orbits is termed a(n)
    A. random approach   D. disjointed approach
    *B. empirical approach   E. bimodal approach
    C. deductive approach

15. The AU is
    A. the radius of the Moon's orbit in the epicycle model
    *B. the average distance of the Earth from the Sun
    C. a measure of the gravitational attraction between two planets
    D. the most recently discovered satellite of Jupiter
    E. none of these

16. One AU is about _____ m.
    A. 5×10^{10} D. 1.5×10^{15}
    B. 1.5×10^{13} E. 2.0×10^{33}
    *C. 1.5×10^{10}

17. If an asteroid has a period of eight years about the Sun, its average distance from the
    Sun is _____ AU. (Hint: Use Kepler's third law.)
    A. 0.65   D. 8
    B. 2   E. none of these
    *C. 4

18. The first known astronomer to use a telescope was
    A. Ptolemy   D. Kepler
    B. Copernicus   *E. Galileo
    C. Brahe

19. Newton's first law states, essentially, that
    A. force equals mass times acceleration
    B. the radius vector from the Sun to a planet sweeps equal areas in equal times
    C. orbits are ellipses with the Sun at one focus
    *D. in the absence of outside forces a body at rest remains at rest
    E. area equals one-half the base times the height

20. Bode's rule proves
   A. that the asteroids were once a planet
   B. that Neptune has been moved since the solar system was formed
   C. that Pluto had to exist
   D. both A and B
   *E. nothing

Problem Involving Optional Mathematical Equations from Other Chapters

*21. (Optional Equation I) Uranus has an equatorial diameter of 51,800 km and is about 19 AU from the Sun. What is the angular diameter of Uranus when it is closest to the Earth? (1 AU = 1.5×10^{10} m)

   A. 39"   D. 40"
   B. 3".8   E. 87"
   *C. 4".0

Essay

1. Describe why the retrograde motion of Mars occurs according to Ptolemy. (See p. 60.)
2. What is Occam's razor? How is it used? (See pp. 61–62.)
3. Decide why the retrograde motion of Mars occurs in the Copernican theory. (See Fig. 3-2.)
4. Give an example of Newton's third law. (See p. 68.)
5. State Newton's three laws of motion. (See p. 68.)
6. State Kepler's three laws. (See pp. 65–66.)
7. State Newton's Universal Law of Gravitation. (See p. 68.)

Word Practice

1. _____ _____ _____ _____ relates the masses, the separation, and the mutual gravitational force between two bodies.

2. The apparent westward drift of Mars at certain times is termed the
   _____ _____ _____ _____.

3. The change in European beliefs from an Earth-centered system to a Sun-centered system is often referred to as the _____ _____.

4. The orbit of a planet is a(n) _____.

5. _____ _____ is best known for providing the observational data of the motion of Mars that Kepler used to derive his three laws.

6. The _____ _____ has the Sun going about the Earth.

7. The Sun and its surrounding worlds are called the _____ _____.

8. The average Earth-Sun separation is one _____ _____.

9. According to Kepler, each planet moves in an ellipse with the Sun at one

   _____.

10. _____ _____ _____ _____ are three laws that are fundamental postulates from which Kepler's three laws can be deduced.

11. _____ _____ tells us to take the physical theory that makes predictions with the fewest assumptions.

12. The _____ _____ of the solar system, perfected by Ptolemy, places the Earth near the center of the solar system.

13. _____ _____ was one of the first astronomers to use a telescope. He wrote his results in Italian.

14. We live on the planet _____.

15. _____ are the passages of an inferior planet directly between the Earth and the Sun.

16. _____ _____ is helpful for remembering the orbital distances of the planets.

17. _____ _____ empirically developed three laws that fit the motions of Mars to an elliptical orbit.

18. The three laws of planetary motion are called _____ _____.

19. _____ _____ was able to deduce Kepler's three laws.

20. The _____ _____ are the ones farther from the Sun than the Earth.

21. The _____ _____ are closer to the Sun than the Earth.

Answers to Word Practice

1. Newton's law of gravitation
2. retrograde motion of Mars
3. Copernican revolution
4. ellipse
5. Tycho Brahe
6. Ptolemaic model
7. solar system
8. astronomical unit
9. focus
10. Newton's laws of motion
11. Occam's razor
12. epicycle model
13. Galileo Galilei
14. Earth
15. transits
16. Bode's rule
17. Johannes Kepler
18. Kepler's laws
19. Isaac Newton
20. superior planets
21. inferior planets

Chapter 4
The Conquest of Gravity and Space

Answers to Problems in the Text

1. Because the height of the Earth's atmosphere is a minuscule fraction of the Earth's radius, we can consider the spaceship to have a circular velocity about equal to that at the surface (8 km/s). Since escape velocity is 11 km/s, the spaceship needs to add only 3 km/s ($11 - 7.91$ km/s if carefully done).

2. From problem 1, we see that to change one orbit into another orbit that passes through the first it is only necessary to change the velocity. In problem 1, a change in velocity of 3 km/s changes a circular orbit into a parabolic (escape) orbit. This change in velocity is often called "delta V," or ΔV, by rocket engineers.

3. First note that

$$1000 \text{ mph} = \frac{1609 \text{ km}}{\text{hour}} = \frac{1609}{60 \times 60} \text{ km/s} = 0.447 \text{ km/s}$$

so the spacecraft has a velocity of 0.447 km/s west to east before launch. (The stars appear to move east to west because the Earth is rotating the opposite way.) Hence, in west-to-east launch, a circular orbit needs only about 7.5 km/s added velocity (making a total of 8 km/s; see problem 1). An east-to-west launch needs only about 8.5 km/s opposite velocity because it must overcome the initial velocity in the west-to-east direction.

4. a. Newton's third law: Action equals reaction. To push yourself toward the dock, you must push the boat in the opposite direction.

   b. If we take the boat as analogous to the rocket, then as you leave the boat you are performing the same function as the rocket exhaust; that is, the particles depart from the rear of the rocket as you are departing from the rear of the boat.

5. Probably either answer can be justified, but the most common should be no. We are now caught in an energy shortage and an inflationary spiral; adding this expenditure would only fuel inflation. (However, you might argue that it would reduce unemployment.) Although the space program takes only a very small percentage of the national budget, it could serve as an obvious "whipping boy." However, if the USSR were to make its first spectacular space launch now, the public might be aroused.

6. (See pp. 84–87.) Again, either side may be argued. Realistically, however, manned flights are very expensive as well as energy- and material-consuming, even though at present rates we are using a small percentage of the budget. It is not clear that the economic situation will allow a political decision to undertake such travel. At present, our manned space program is equivalent to building an airplane to fly one passenger from New York to San Francisco and then throwing the airplane away. The space shuttle and other advances may reduce the cost enormously. There is no question from an intellectual point of view that such a development of space travel would be a great advance. It might also lead to revolutionary technologies in space (such as energy-producing satellites). We might ask if our society can exist for the next hundred years without developing manned interplanetary travel. Perhaps in the next hundred years we can solve our problems on the Earth sufficiently that space travel can become a symbol of great social progress.

7. Some students may make this a difficult problem if they realize that the form of Newtonian gravitational law given is valid only if we are outside a spherical body. In this problem, we are looking at effects that depend upon the shape of the Earth. The intent is to take the force law and see the effect of distance from the center. Let M be the Earth's mass, m your mass, r the sea level radius of the Earth, and h the altitude above (+) or below (–) sea level. Then if w is the weight,
$$w(h) = GMm/(r+h)^2.$$
If we look at the ratio $w(h)/w$ (sea level), we get
$$w(h)/w \text{ (sea level)} = 1/(1 + h/r)^2.$$
In the Dead Sea, $h/r = -0.000062$, so
$$w(DS) = 1.0001\ w \text{ (sea level)}.$$
On Mt. Everest, $h/r = 0.0014$, so
$$w(ME) = 0.997\ w \text{ (sea level)}.$$
Assuming a weight of 150 lbs, this is a gain of about 1/4 oz in the Dead Sea and a loss of 7 oz on Mt. Everest.

8. Use weight = w = GMm/r^2 and the notations p for the planet and E for the Earth.
   a. So $w(p)/w(e)$ $= (M(p)/M(e)) \times (r(e)/r(p))^2$
   $= 0.11/(1/.53)(1/.53) = 0.031.$
   b. The mass ratio and size ratio for Mars/Earth from Table 8-1 are the same as used in part a, so the value of weight must be the same.

9. At the Moon's distance, as the text tells us, the velocity for a circular orbit is 1 km/s. Now the escape velocity is $\sqrt{2}$ times the circular velocity, so escape velocity is 1.414 km/s. Hence, he or she would need to increase the velocity by only 0.414 km/s.

10. $V_{circ} = \sqrt{GM/R}$ $G = 6.67 \times 10^{-11}$ Nm^2/kg^2
    M = mass of Earth = 5.98×10^{24} kg
    R = 42,500 km = 4.25×10^7 m
    $= 6.66 \times R_\oplus = (2.58)^2 R_\oplus$

Putting the numbers into the formula:

$$V_{circ} = \left(\frac{6.67 \times 10^{-11} \times 5.98 \times 10^{24}}{4.25 \times 10^7}\right)^{1/2} \text{m/s} = 3.0 \text{ km/s}.$$

Alternatively, because $R = (2.58)^2 R_\oplus$

$$V_{circ} = \sqrt{\frac{GM_\oplus}{(2.58)^2 R_\oplus}} = \frac{1}{2.58}\sqrt{\frac{GM_\oplus}{R_\oplus}} = \frac{7.91 \times 10^3}{2.58} \text{ m/s}$$
$$= 3.1 \text{ km/s}.$$

A circle of radius 42,000 km has a circumference of 2.67×10^5 km. The time for one orbit

$$T = \text{circumference}/V_{circ}$$

so $T = \dfrac{2.67 \times 10^5 \text{ km}}{3.1 \text{ km/s}} = 8.6 \times 10^4$ s.

In one day there are $24 \times 60 \times 60 = 8.64 \times 10^4$ s. These are nearly the same, so the satellite has almost exactly a 24-hour period. The difference is mostly due to our approximations and round-off error.

Sample Test Questions

True-False

1. Isaac Newton discovered the principle of gravitational attraction in the 1600s. T
2. The gravitational force follows a square law. F
3. If the distance between two gravitating bodies doubles, the force drops by a factor of four. T
4. If the distance between two gravitating bodies doubles, the force increases by a factor of four. F
5. The circular velocity is the speed an object must move parallel to the surface of a body in order to escape from it. F
6. If a spaceship is launched from the Earth with a higher speed than the escape velocity, it moves on a hyperbola. T
7. Newton's third law of motion states that for every action there is an equal reaction in the same direction. F
8. The first artificial satellite, Sputnik I, was launched in 1958. F
9. The first man in space was an American, Alan Shepard. F
10. The decision by President Kennedy to put a man on the Moon in the 1960s was more a political decision than a scientific decision. T
11. Space flight has had no practical results. F

Multiple Choice

1. Newton's law of gravitational force is one of his three laws of motion.
   - A. correct
   - *B. wrong: This is a force law, not a law of motion.
   - C. misleading: Newton had laws of motion.
   - D. wrong: Both B and C corrections are needed.

2. If r is the distance between two gravitating bodies, then the force produced by one on the other is
   - A. proportional to r
   - B. proportional to r^2
   - C. inversely proportional to r
   - *D. inversely proportional to $r2$
   - E. independent of r

30

3. If we have a body of mass A and a body of mass B separated by a distance D, then the force F that one produces on the other is proportional to
    A. AD/B^2 D. ABD
    B. BD/A^2 E. $A/(BD)$
  *C. AB/D^2

4. The weight of the body is
    A. the same as its mass
  *B. the gravitational force by which the Earth pulls a body
    C. the same on all planets
    D. both B and C
    E. a measure of the amount of matter it contains

5. The mass of a body is
    A. the gravitational force by which the Earth pulls a body
    B. the same on all planets
    C. both A and B $b + D$
  *D. a measure of the amount of matter it contains
    E. none of the above

6. Newton's first law of motion is
    A. for every action there is an equal and opposite reaction
    B. the gravitational force is inversely proportional to the distance
  *C. a projectile keeps moving forward unless a modifying force acts on it
    D. the gravitational force is proportional to the mass
    E. none of the above

7. Newton's third law of motion is
  *A. for every action there is an equal and opposite reaction
    B. the gravitational force is inversely proportional to the distance
    C. a projectile keeps moving forward unless a modifying force acts on it
    D. the gravitational force is proportional to the mass
    E. none of the above

8. The circular velocity at the Earth's surface is about _____ km/s.
    A. 3   D. 11
    B. 5   E. 1
  *C. 8

9. An orbiting body's point of closest approach to the Earth is called its
    A. ellipse   *D. perigee
    B. eclipse   E. parabola
    C. apogee

31

10. The speed at which an object can escape from the Earth forever is termed the
_____ velocity.
  A. perigee
  B. hyperbolic
  C. thrust
  D. apogee
  *E. escape

11. The closed orbit that a satellite might have about the Earth is called
  *A. an ellipse
  B. a parabola
  C. a hyperbola
  D an eclipse
  E. a perigee

12. The first artificial Earth satellite was launched in _____ by
_____.
  A. 1887/France
  B. 1957/USA
  *C. 1957/USSR
  D. 1959/USA
  E. 1959/USSR

13. The United States launched its first successful artificial Earth satellite in
_____.
  A. 1955
  B. 1956
  *C. 1957
  D. 1958
  E. 1959

14. The first human being in space was _____ in _____.
  A. Jules Verne/1957
  B. Yuri Gagarin/1961
  *C. Yuri Gagarin/1957
  D. Alan Shepard/1957
  E. Alan Shepard/1961

15. The _____ are doughnut-shaped zones of energetic particles surrounding the
Earth.
  *A. Van Allen belts
  B. Von Braun zones
  C. Jules Verne regions
  D. Newtonian hyperbolas
  E. Shepard zones

16. The NASA budget in the years 1959–1969 was 25% of the total national budget.
  A. correct
  B. no: 75%
  C. no: 1%
  *D. no: 2.5%
  E. NASA was not established
     until 1969.

*17. If G is Newton's gravitational constant, M the mass of a central body, and R the
distance of an orbiter from the center of the body, then the circular velocity of the
orbiter is
  A. GM/R
  B. $R/(GM)$
  C. $\sqrt{RGM}$
  D. $\sqrt{R/GM}$
  *E. $\sqrt{GM/R}$

*18. If the circular velocity at the Earth's surface is 7.91×10^5 cm/s, at four times the Earth's radius the circular velocity is about _____ $\times 10^5$ cm/s.
    *A. 3.96   D. 1.0
     B. 7.91   E. 0.53
     C. 14.2

*19. If the Earth's mass were doubled, the circular velocity would be
     A. doubled   D. divided by $\sqrt{2}$
     B. halved   E. none of these
   *C. multiplied by $\sqrt{2}$

*20. The circular velocity at the surface of Jupiter is 42.5 km/s. The approximate escape velocity is
     A. 11 km/s   *D. 60 km/s
     B. 30 km/s   E. 85 km/s
     C. 42.5 km/s

*21. If the radius of Venus is taken as equal to the Earth's and its mass is 0.82 times that of the Earth, the circular velocity at the surface of Venus is about _____ times that of the Earth
     A. 0.82   D. 1.1
     B. 1.22   E. 1.0
   *C. 0.905

*22. The escape velocity for a planet with twice the Earth's mass and the same radius as the Earth would be _____ times the escape velocity from the Earth.
     A. 2   *D. $\sqrt{2}$
     B. $1/\sqrt{2}$ E. none of these
     C. $2\sqrt{2}$

23. Newton's law of gravitation for two bodies of masses M_1 and M_2 separated by distance D with G a constant is
   *A. $F = GM_1M_2/D^2$ D. $F = GM_1M_2/D$
     B. $F = FM_1M_2D^2$ E. none of these
     C. $F = GM_1M_2D$

Essay

  1. Describe three direct practical results of the space program. (See pp. 85–87. Some main points: weather predictions, communications, photography for mapping, navigational uses.)
  2. Describe two intangible results of the space program. (See p. 87. Some main points: management techniques with technology, cosmic perspective, human survival.)

*3. A small artificial satellite of 50 cm radius is put in Earth orbit with a perigee 200 kilometers above the Earth's surface. What is its angular diameter at perigee? What if the radius were 500 m? In either case, could the satellite be resolved by the naked eye? (See Chapters 2 and 4. Answers: 0".5; 500"; yes, the 500 m one should be resolvable.)

4. State Newton's first law of motion. (See p. 75.)

5. Why is the gravitational force law not considered a law of motion? (See p. 75. Laws of motion tell how bodies behave in general; the gravitational force law gives the force if gravity is involved.)

Word Practice

1. Newton's _____ _____ _____ _____ has action equaling reaction.

*2. The first American in space was _____ _____.

3. The _____ _____ of the Earth is 11 km/s.

4. Newton's _____ _____ _____ gives the force a massive body produces on another body.

5. The _____ _____ is the speed necessary for circular orbit.

6. The closed orbit of one body about another is called a(n) _____.

7. A(n) _____ is the orbit resulting from speeds above escape velocity.

*8. One of the practical results of the space program was _____ satellites.

9. The _____ _____ _____ are doughnut-shaped zones of energetic atomic particles about the Earth.

10. _____ is the force exerted by exhaust gases of a rocket.

11. The _____ is a satellite's closest approach to the Earth.

*12. The first Russian satellites were named _____.

13. _____ was responsible for formulating the law of gravitation.

*14. The first man in space was _____ _____.

*15. NASA's budget was _____% of the U.S. national budget during 1959–1969.

16. A satellite's farthest departure from the Earth is its _____.

17. _____ is a measure of the amount of matter in a body.

18. An orbit with exact escape velocity is a(n) _____.

19. The law of gravitation is a(n) _____ _____ _____.

*20. _____ from satellites can help with mapping.

34

21. Newton's _____ law of motion states that an object continues in constant motion unless acted upon by an outside force.

Answers to Word Practice

1. third law of motion
2. Alan Shepard
3. escape velocity
4. law of gravitation
5. circular velocity
6. ellipse
7. hyperbola
8. communications
9. Van Allen belts
10. thrust
11. perigee

12. Sputnik
13. Newton
14. Yuri Gagarin
15. 2.5
16. apogee
17. mass
18. parabola
19. inverse square law
20. photography
21. first

Chapter 5
Light and the Spectrum: Messages from Space

Answers to Problems in the Text

1. Since the energy per photon is larger for shorter wavelengths, we would expect it to be easiest to measure blue light and hardest to measure infrared light.

2. Water vapor and clouds tend to absorb and reradiate or to reflect heat. Thus, on a warm, cloudy night, less heat escapes than on a dry, clear night.

3. Since only the absorption bands of CO_2 produce featureless photos, we expect the atmosphere of the planet to be nearly pure CO_2.

4. Our atmosphere absorbs most of the ultraviolet and infrared radiation hitting it, while most of the radio waves pass through. Thus, on the Earth's surface most of the radio waves can be detected, while UV and IR telescopes receive little radiation. To study UV and IR, we need to get the telescopes above the Earth's atmosphere.

5. Two binoculars of 50-mm aperture, but one of 4× and the other of 10× magnification.
   a. Magnifying power = (angular size with aid)/(angular size without aid), so the 10× gives a larger apparent image size.
   b. Two cases—for stars, since both binoculars have the same aperture, the images will be equally bright. If we look at extended objects (Moon, planets), then since the 10× pair produces a bigger image, that image must be dimmer than the image produced by the 4× (that is, the same amount of light is spread over a larger area with the 10×).
   c. The lower power will be easier to hold still. The 10× is about at the limit for hand-held instruments.

6. For a camera, the lens is the objective and there is no eyepiece. The farther an image forms from a lens, the larger the image. For an astronomical object, the image forms at the focal length. We also expect that the larger the magnifying power the smaller the field of view, so
   a. the 90-mm lens
   b. the 24-mm lens
   c. the 90-mm lens

7. Emission lines are produced when an electron emits a photon and goes to a lower energy level.
   a. The ground state is the lowest energy level, so no photon can be emitted.
   b. Near hot stars, the atoms in a nebula are in high energy levels (the atom is excited). This occurs because the hot atoms often collide very hard and some of this energy of motion is transferred to the electronic structure. Thus, emission can occur as electrons are in their lowest states, and no emission can occur from electronic excitation.

8. The curve in Fig. 5-14 peaks at about 2×10^{-5} m. Use Wien's law
   $$T = 0.0029/L,$$
   where T is the Kelvin temperature and L is the wavelength of the peak of the curve. So
   $$T = 0.0029/(2 \times 10^{-5}) = (2.9/2) \times 10^{-3+5} = 1.45 \times 10^2 = 145 \text{ K.}$$

9. Use Wien's law. The peak is at $L > 10 \times 10^{-6}$ m. If the peak were at $10 \times 10^{-6} = 10^{-5}$ m, the temperature would be $T = 0.0029/10^{-5} = 200$ K. The true L is larger, so $T < 200$ K.

10. The book is at a temperature of about $27°$ C $= 300$ K. (A good rule of thumb is room temperature $= 300$ K.) Use Wien's law,
    $$L = (2.9 \times 10^{-3}/300) \text{ m} = 9.7 \times 10^{-6} \text{ m.}$$

11. a. Magnification = (focal length of objective)/(focal length of eyepiece). This is a ratio, so we can leave both in inches and get
    $$M = 70(1/2) = 140.$$
    b. The image brightness depends upon the aperture, which is information that was not given, so we cannot tell.
    c. Use the other definition of magnifying power:
    $$\text{magnifying power} = \frac{\text{angular size with aid}}{\text{angular size without aid}}$$
    $140 = $ size/ $(1/2)$
    size $= 70°$.
    d. If the friend wants the aperture size, we do not have enough information. If what the friend wants is the magnifying power, we can give the answer, 140×.

12. The larger the magnifying power, the smaller the field of view and the fainter an extended object appears.
    a. The 150× since Mars is small (20"). Thus,
    $$20" \times 150 = 3000" = 50' = .83°,$$
    while for the 30×,
    $$20" \times 30 = 10',$$
    which is too small to be useful.
    b. The 30× since the image with the 150× would be extremely large and hence too faint (also, so large that only a small portion wold be visible in the field of view).
    image size (30×) $= 3 \times 30 = 90°$
    image size (150×) $= 3 \times 150 = 450°$.

Sample Test Questions

True-False

1. In a reflecting telescope, the objective is a mirror. T
2. All radio telescopes are refractors. F
3. One reason useful magnification is limited is that at too high a magnification the image is too bright. F

4. The maximum useful magnifying power for most telescopes is about 20. F
5. The light-gathering power of a telescope increases as the aperture is enlarged. T
6. Faint objects can be seen better on photographs than with the eye. T
*7. The hotter an object, the bluer the radiation it emits. T
*8. The shortest wavelength of radiation that can be detected by the normal eye is about 4 × 10^{-7} m. T
*9. The shortest wavelength of radiation that can be detected by the normal eye is about 6.5 × 10^{-7} m. F
*10. Radiation with a wavelength longer than visible light is termed ultraviolet. F
11. The study of the colors of light emitted by objects is called spectroscopy. T
12. The shortest wavelength of visible radiation is violet. T
13. The discrete units of light are called photons. T
14. The array of colors in a light beam, ordered according to wavelength, is called white light. F
15. A glow consisting of a mixture of all colors is called an emission line. F
16. A molecule emits an absorption line. F

Multiple Choice

1. The dark lines in the spectrum are known as _____ lines.
   *A. absorption
   B. continuous
   C. missing
   D. hot
   E. none of the above

*2. Wien's law states:
   A. The cooler an object, the bluer the radiation it emits.
   *B. The hotter an object, the bluer the radiation it emits.
   C. The hotter an object, the redder the radiation it emits.
   D. Temperature is proportional to emitted energy.
   E. Force is equal to mass times acceleration.

*3. The eye detects radiation with a wavelength of _____ to _____ m.
   *A. 4 × 10^{-7}/6.5 × 10^{-7}
   B. 4 × 10^{-10}/6.5 × 10^{-10}
   C. 4 × 10^{-7}/9 × 10^{-7}
   D. 4 × 10^{-10}/9 × 10^{-5}
   E. 2 × 10^{-9}/7 × 10^{-3}

*4. Light of a wavelength of about 6.5 × 10^{-7} m is
   A. deep blue
   *B. deep red
   C. yellow
   D. light green
   E. ultraviolet

*5. Light with a wavelength of 3 × 10^{-7} m is
   A. visible
   B. infrared
   *C. ultraviolet
   D. radio
   E. green

*6. A star has a temperature of 7000 K. The maximum amount of radiation is at a
    wavelength of about _____ m.
    *A. 4.1×10^{-7} D. 6.5×10^{-5}
     B. 2.0×10^{-5} E. not near any of these
     C. 3×10^{-7}

*7. A star has a temperature of 4500 K. What color is it?
     A. red   D. yellow
    *B. blue   E. white
     C. green

*8. Which of the following is the relation between the wavelength, W (in m), at which
    the maximum amount of radiation is emitted, and the Kelvin temperature T ?
     A. $W = 0.00290T$ *D. $W = 0.00290/T$
     B. $W = 3/T$ E. none of these
     C. $W = 2.90/T$

 9. In a refracting telescope, the objective is a
    *A. lens   D. grating
     B. mirror   E. none of these
     C. prism

10. The distance between the image that a distant object forms and the main lens or
    mirror is the _____ of the lens.
     A. objective   D. magnifying power
     B. refractor   E. secondary
    *C. focal length

A — mirror, lens, parabolic mirror
B — mirror, parabolic mirror, lens
C — lens, lens

    D. both A and B   E. none of these

11. In the diagram above, which of these is (are) a reflecting telescope? D

12. In the diagram above, which of these is (are) a refracting telescope? C

39

13. You are using a 30-cm diameter telescope with a 120-cm focal length and an eyepiece with a focal length of 6 mm. What is the magnifying power?
    A. 20
    *B. 200
    C. 420
    D. 600
    E. none of these

14. Using a telescope with a resolution of 3", can two stars be resolved if they are separated by (1) 1"? (2) 9"?
    A. yes for (1), no for (2)
    B. yes for (1), yes for (2)
    C. no for (1), no for (2)
    *D. no for (1), yes for (2)
    E. not enough information to tell

15. As a practical matter, resolution is limited by
    A. diffraction
    *B. the atmosphere
    C. the Sun
    D. present quality of lens grinding
    E. spectroscopy

16. What is the normal maximum usable magnifying power of a 30-cm aperture telescope with a focal length of 200 cm?
    A. 10
    B. 300
    C. 4000
    D. 500
    *E. 600

17. Which has the most light-gathering power?
    *A. a telescope with a 30-cm aperture and a 100-cm focal length
    B. a telescope with a 25-cm aperture and a 200-cm focal length
    C. a telescope with a 25-cm aperture and a 100-cm focal length
    D. a telescope with a 10-cm aperture and a 100-cm focal length
    E. both B and C

18. Which is best for getting a sharp image of Mars?
    A. the eye alone
    *B. using the telescope to view (telescope + eye)
    C. using the telescope to photograph (telescope + camera)
    D. a radio telescope
    E. a crystal ball

19. The process of combining the information from two or more telescopes to increase resolution is called
    A. absorption
    B. digitizing
    C. false color process
    *D. interferometry
    E. spectrophotometry

40

Essay

1. Draw a refracting telescope and the light rays through it from a distant "star." Label objective, eyepiece, and both focal lengths. (See Fig. 5-20.)
2. Draw a reflecting telescope and the light rays through it from a distant "star." Label objective, eyepiece, and any other relative part. (See Fig. 5-20.)
3. What are the three effects that limit the usable magnifying power of a telescope? (See pp. 109–110.)
4. Discuss the advantages of the eye versus a camera when using a telescope. (See p. 111.)
5. Distinguish between an absorption line and an absorption band. Under what conditions does each occur? (See pp. 102–103.)
6. Discuss the differences in the wave behavior of light and its particle aspect. (See pp. 93–94.)
7. Give three ways that telescopic images can be stored and modified. (See pp. 112–113.)
8. Describe interferometry. (See p. 115.)
9. Discuss light pollution and how it affects astronomical observation. (See p. 115.)

Word Practice

*1. The _____ is a small lens system used to magnify the image formed by the objective.

*2. The eyepiece is used to magnify the _____ formed by the objective.

3. The _____ of a telescope is the diameter of the objective.

4. The _____ _____ of a telescope is determined solely by the aperture size.

5. The main lens or mirror of a telescope is called the _____.

6. _____ telescopes use lenses and mirrors together for their objective.

7. A(n) _____ uses a mirror as its objective.

*8. The _____ mirror present in some telescopes is primarily to beam the light rays to a location for easier access.

*9. A telescope with a long-focal-length eyepiece will have a(n) _____ power compared to the same telescope with a short-focal-length eyepiece.

*10. The shorter the focal length of the _____,the larger the magnifying power.

*11. There is a limit to a normal useful _____ of a telescope. This limit is about 20 power per cm of aperture.

*12. The _____ _____ of a telescope is the angular size of the object as seen through the telescope divided by its angular size without the telescope.

13. A(n) _____ uses a lens as its objective.

*14. A(n) _____ reflector is a telescope that uses a secondary mirror to direct the light out the side of the telescope.

*15. A(n) _____ telescope is designed to be looked through.

*16. A(n) _____ is a small telescope attached to a larger telescope. It is used to guide the larger telescope when making photographs.

17. The _____ length of a lens or mirror is the distance from it that the image of a distant object forms.

18. The _____ is a measure of the clarity of the atmosphere.

19. The process of splitting light into wavelengths so the intensity of each wavelength can be measured by a photomultiplier tube is _____.

20. The process of measuring the amount of light from an object, usually with electronic devices, is called _____.

21. The spacing between crests of a wave is the _____.

22. One of the _____-like properties of light is its localization.

23. The period between pulsations of a light wave is its _____.

24. The spreading of light in all directions from a source is one of the _____-like properties of light.

25. _____ light has a wavelength shorter than visible light.

26. _____ light has a wavelength longer than visible light.

27. The bending of light as it passes an edge is called _____.

28. The arrangement of light by colors is called the _____.

29. _____ is the measure of the average motions of the molecules of a substance.

30. The process of an atom emitting a photon is called _____.

31. The broad emission feature from a molecule is called a(n) _____ _____.

32. Radiation not related to the thermal motions of atomic particles is called
_____ radiation.

33. The _____ is a measure of how many times bigger the angular size of an object appears through a telescope.

34. The _____ space telescope will be the first large telescope in space.

35. The _____ of a telescope is a measure of the detail that can be seen through the telescope.

36. The larger the _____ _____ of a telescope, the brighter the image it produces.

37. The process of combining signals from separate telescopes to improve resolution is called _____.

Answers to Word Practice

1. eyepiece
2. image
3. aperture
4. light-gathering power
5. objective
6. compound
7. reflector
8. secondary
9. smaller
10. eyepiece
11. power or magnifying power
12. magnifying power
13. refractor
14. Newtonian
15. visual
16. finder
17. focal
18. seeing
19. spectrophotometry
20. photometry
21. wavelength
22. particle
23. frequency
24. wave
25. ultraviolet
26. infrared
27. diffraction
28. spectrum
29. temperature
30. emission
31. emission band
32. nonthermal
33. magnification
34. Hubble
35. resolution
36. light-gathering power
37. interferometry

Answers to Problems in the Text

1. The answer depends on the student's home location, which each student might discuss. Sources of information are local geologists and insurance agents. Insurance underwriters have an overly pessimistic map, but it is useful.

2. The half-life of U-235 is 7.1×10^8 years. (This figure is not in the text.) With each half-life, the amount of U-235 left decreases by half. In other words, after one half-life there is half as much. After two half-lives there is $1/2 \times 1/2 = 1/4$ as much. After three half-lives there is $1/2 \times 1/2 \times 1/2 = 1/8$ as much. So the rock sample is 2.1×10^9 years old.

3. The indicated surface features show that the Moon is much less active than the Earth, so there is much less "earthquake" activity on the Moon than on the Earth. On the Earth, earthquakes are usually the result of plate tectonics, a process that has not occurred on the Moon.

4. We take the Earth's age as 4.6×10^9 years. Then if the genus *Homo* is 3 million years = 3×10^6 years old, the fractional duration of its existence so far is

$$(3 \times 10^6) / (4.6 \times 10^9) \quad = \quad (3/4.6) \times 10^6/10^9$$
$$= (0.65) \times 10^{-3}$$
$$= 6.5 \times 10^{-4} \, .$$

Multiply this by 100 to get percent, which gives $6.5 \times 10^{-2}\% = 0.065\%$—that is, much less than 1/10 of a percent.

5. The half-life of U-238 is 4.51 eons. The Earth is 4.6 eons old, so about one-half of the original U-238 is left. Because heat production from U-238 decay depends upon the total number of atoms decaying, the original heat output was higher than now.

6. a. Most of the hydrogen was lost to outer space. If the Earth had been much more massive, such as Jupiter is, the hydrogen would have been retained.
   b. The nitrogen released by the breakup of NH_3 provided much of the nitrogen in our atmosphere.
   c. Volcanoes added carbon dioxide and water vapor. The water vapor produced the oceans, while the carbon dioxide dissolved in the water and reacted with the Earth's surface to form carbonate rocks.

7. This question asks, in effect: What would happen to the Earth as it is now if it were suddenly placed at either the orbital distance of Venus from the Sun or the orbital distance of Mars from the Sun?

Case 1 (if the Earth were closer to the Sun): Temperature would be above 100° C; most of the oceans would evaporate, and life would cease.

Case 2 (if the Earth were farther from the Sun): Temperature of the Earth's surface would be below 0° C; the oceans would freeze over, eventually destroying all life in them.

As extra information (beyond what the question asked), we note that in either case, since life has ceased, the free oxygen and nitrogen in the atmosphere would eventually disappear, leaving the Earth with a carbon dioxide atmosphere (and water vapor in Case 1).

Advanced Problems

8. a. Surface temperature is 20° C, boiling water is 100° C, so we want an 80° C temperature change. At 20° C/cm, we would need to go 80° C/(20° C/km) = 4 km. This is roughly the depth of the deepest mines. In reality, this calculation fails because within a kilometer of the surface the gradient falls from 20° C/km to a lower figure. The actual depth would be greater than 10 km; however, deep mines are notably hot.

    b. *Economic consequences*
      1) reduction—although not elimination—of energy crisis
      2) elimination of most coal mining or its conversion to a hydrocarbon source in industry
      3) probable reduced air pollution, but increased thermal pollution
      4) shift in economic power of various industries

    c. *Political consequences*
      1) reduction of political power of OPEC countries
      2) possible rapid industrial development of Third World countries because all countries would be independent of foreign energy sources

9. a. We need the angular size of the Earth as seen from Mars.

Use $\dfrac{\alpha''}{206,265} = d/D$

where $d = 1.28 \times 10^4$ km $= 1.28 \times 10^7$ m
     $D = 6 \times 10^7$ km $= 6 \times 10^{10}$ m

so $\alpha_\oplus = 2.06 \times 10^5 \times d/D$

     $= (2.06 \times 1.28/6) \times (10^5 \times 10^7/10^{10}) = 43''$

while the Sun is

    $\alpha_\odot = 1/3° \times (60'/1°) \times (60''/1') = 1200''.$

Hence, the Sun is larger in angular diameter than the Earth, so eclipses would be annular at best. The Earth would be a very tiny dot against the Sun.

    b. No. The disk of Earth is 43", which is smaller than the resolution of the naked eye (120").

Sample Test Questions

True-False

1. The Earth is perfectly regular in its rotation. F
2. An eastward or westward deflection of moving air is called a Newtonian drift. F
3. The Earth is estimated to be 4.6×10^8 years old. F
4. A radioactive atom is a stable atom that does not change. F

5. Ages of rocks can be dated by counting relative amounts of radioactive decay products. T
6. The oldest known Earth rocks are found on circular regions called continental plates. F
7. The Earth is much younger than the Moon and most of the other planets. F
8. The center of the Earth's core is liquid. F
9. Earthquakes occur in the Earth's core. F
10. The crust is the outermost shell of the Earth. T
11. The process that separates the elements in the Earth is termed differentiation. T
12. Melting of the Earth that causes gases to be released is termed convection. F
13. Heat transmission by the motion of materials is termed convection. T
14. The process of crustal disruption by flow of materials is termed plate tectonics. T
15. Mountains are caused by the Earth's shriveling. F

Multiple Choice

1. The outermost layer of the Earth is the
   A. core
   *B. crust
   C. mantle
   D. asthenosphere
   E. epicycle

2. A body moving north in the Northern Hemisphere is deflected to the _____ by the _____ drift.
   A. east, Foucault
   B. west, Coriolis
   *C. east, Coriolis
   D. west, Foucault
   E. north, Smith

3. On the Earth, the effect that causes a deflection in the path of a moving body is the _____ effect.
   *A. Coriolis
   B. Foucault
   C. Newtonian
   D. deflecting
   E. Hartmann

4. The best current estimate of the Earth's age is _____ years.
   A. 5,982
   B. 4.6×10^3
   C. 4.6×10^6
   *D. 4.6×10^9
   E. 1.8×10^{10}

5. When an original radioactive element decays, the new element formed is termed the
   A. radioactive atom
   B. parent isotope
   *C. daughter isotope
   D. residue
   E. half-life

6. Suppose a radioactive element has a half-life of 300 years. After 600 years, how much of the original amount will be left?
   - A. 1/2
   - *B. 1/4
   - C. 1/8
   - D. 1/16
   - E. twice

7. The technique of radioisotopic dating measures the time since the rocks being studied
   - A. were formed from a dust and gas cloud
   - B. were dug up
   - C. were last exposed to sunlight
   - *D. were solidified from an earlier molten material
   - E. none of these

8. Flat, circular regions, found on several continents, that yield the oldest known rocks are the
   - A. circular field
   - B. massive crumplings
   - C. continental disks
   - D. old schliefs
   - *E. continental shields

9. Sound waves traveling through the Earth are termed _____ waves.
   - A. earth traveling
   - *B. seismic
   - C. magnetic
   - D. geologic
   - E. fault

Consider the sketch of the Earth:

10. In this diagram, the core is denoted by the letter __A__.

11. In this diagram, the mantle is denoted by the letter __B__.

12. In this diagram, the crust is denoted by the letter __C__.

Earth

13. The Earth's core is primarily composed of
   - A. lead
   - B. silicates
   - C. copper
   - *D. nickel-iron
   - E. iron

14. The Earth's magnetic field is believed to be produced by liquid metal circulating in the Earth's
   - A. crust
   - B. mantle
   - *C. outer core
   - D. inner core
   - E. misleading: This is not the cause of the Earth's magnetic field.

47

15. Mixtures of crystalline compounds called _____ combine into groups to form various kinds of _____.
    A. elements/compounds   *D. minerals/rocks
    B. atoms/molecules   E. plates/continents
    C. rocks/minerals

16. The outermost shell of the Earth is the
    *A. crust   D. declination
    B. mantle   E. none of these
    C. core

17. A dark gray volcanic rock is called
    A. silicate   D. stony-iron
    B. granitic   E. none of these
    *C. basaltic

18. The group of processes that have separated some groups of chemical elements from others is termed
    A. separation   D. conduction
    B. drift   *E. none of these
    C. isotope

19. When gases are released by heating, melting, or partial melting of a material, the process is termed
    A. differentiation   D. radioactive decay
    B. conduction   *E. outgassing
    C. radiation

20. _____ is the flow of heat through a material by the mechanical agitation of adjacent molecules in the material.
    A. outgassing   D. radiation
    *B. conduction   E. none of these
    C. convection

21. The eruption of molten materials from a planet's interior onto its surface is
    A. radiation   D. faulting
    B. impact cratering   E. none of these
    *C. volcanism

22. Which of these processes affect(s) the Earth's surface?
    A. impact cratering   D. B and C, but not A
    B. volcanism   *E. A, B, and C
    C. erosion

23. All processes of fracturing and movement of the Earth's crust are called
_____. But _____ refers to the effects of a sudden cracking.
   A. motion/sudden
   B. motion/earthquake
   *C. tectonics/earthquake
   D. earthquake/tectonics
   E. none of the above

24. The continents were all part of a single massive continent called Pangaea about 7
billion years ago. Convection currents caused the continent to break up into pieces,
setting off continental drifts.
   A. correct in all aspects
   B. wrong: The continents never were together.
   *C. wrong: It was 300 million years ago.
   D. wrong: The breakup was caused by radiation, not convection.
   E. wrong: Both changes C and D are needed to correct the statement.

Essay

1. Describe the evolution of scientific thought about the Earth's age. (See pp. 121–124.
Some main points that might be considered:
Archbishop Ussher, in 1650, estimated 4004 B.C.

late 1600s:	Recognition that sediments gave earlier dates.
1700s:	Fossils and strata give earlier dates.
1500s–1800s:	Belief that mountains form by shriveling.
1900s:	Development of radioactive dating; ages of the Earth's surface rocks estimated.
mid-twentieth century:	Dating of meteorites and Moon samples; age of solar system.)

2. Describe how earthquakes are used to determine the Earth's structure. A drawing
might help. (See p. 124. Some main points: Earthquakes produce seismic waves
that are bent and reflected by structure. Use a figure similar to Fig. 6-21 to show
delay of waves.)

3. Discuss the evidence for continental drift. (See pp. 127–130. Some main points:
   1620: Francis Bacon noticed fit.
   1922: A. L. Wegener proposed drifting earthquake patterns and mountain
   building; lowering sea floor at spreading lines.)

4. Let the history of the Earth be one day long. When did fish appear? Mammals?
Human beings? (See pp. 137–139.)

5. Discuss the evidence supporting the asteroidal cause of the Cretaceous-Tertiary
extinctions. (See pp. 136–137.)

6. Discuss the theory of nuclear winter. (See p. 136.)

Word Practice

1. A(n) _____ atom is an unstable atom that decays into simpler atom(s).

2. The time it takes half of a radioactive sample to decay is that element's

   _____.

3. The possibility of major weather changes caused by atomic war is termed the theory

   of _____ _____.

4. _____ is a group of processes separating chemical elements.

5. A(n) _____ _____ is a new atom formed by radioactive decay.

6 4.6 billion years is the estimated _____ _____ _____

   _____.

7. _____ _____ travel through the Earth.

*8. The materials in the Earth and the way they are arranged are the Earth's

   _____.

9. Dark gray volcanic rock is called _____.

*10. The Earth's outer core is _____ nickel-iron.

*11. _____ is a release of gases by melting of underground material.

12. Different _____ of an element have the same number of protons but a

   different number of neutrons.

13. The part of the Earth directly beneath the crust is the _____.

14. The crust of the Earth is composed of distinct _____.

15. _____ is the process of fracturing and movement of the Earth's crust.

16. Molten rock that reaches the surface is called _____.

17. The original atom before radioactive decay is called the _____ isotope.

18. The eruption of molten materials from a planet's interior is called

   _____.

*19. _____ is a heat transport process by motions of materials.

20. A(n) _____ is the result of motion of rock units with respect to each other.

*21. If all minerals are melted, we have _____ _____.

22. _____ is energy carried by electromagnetic waves.

23. _____ is a process removing and transporting materials across a planet's

   surface.

24. Liquid metal circulating in the outer part of the Earth's core sets up electric currents that create the Earth's _____ _____.

25. _____ _____ is the eastward and westward deflection of a moving object due to the Earth's rotation.

26. _____ is a silicate-rich rock of lower density than basalt.

27. _____ _____ involves estimating ages of rocks from relative amounts of radioactive materials.

*28. _____ are crystalline compounds formed from elements.

*29. _____ are mixtures of minerals.

30. Processes that deposit and accumulate materials on a planet's surface are called _____.

*31. A sudden cracking of the Earth's crust is called a(n) _____.

*32. When _____ _____ occurs, part of the material flows away, leaving solids behind.

*33. _____ is the flow of heat through a material by molecular vibrations.

34. _____ _____ are circular depressions created by interplanetary debris colliding with the Earth.

*35. Convection currents in the mantle cause the crustal plates to move, producing _____ _____.

36. Stable, ancient regions are called continental _____.

37. The center part of the Earth is its _____.

38. The _____ is the most rigid part of the Earth.

39. The _____ is a fluid and plasticlike zone.

40. A physical property associated with a region of space is a(n) _____.

41. The _____ extinctions were probably caused by a large meteorite impact.

1. radioactive
2. half-life
3. nuclear winter
4. differentiation
5. daughter isotope
6. age of the Earth
7. seismic waves
8. structure
9. basalt
10. liquid
11. outgassing
12. isotopes
13. mantle
14. plates
15. tectonics
16. lava
17. parent
18. volcanism
19. convection
20. fault
21. complete melting
22. radiation
23. erosion
24. magnetic field
25. Coriolis drift
26. granite
27. radioisotopic dating
28. minerals
29. rocks
30. deposition
31. earthquake
32. partial melting
33. conduction
34. impact craters
35. continental drift
36. shields
37. core
38. lithosphere
39. asthenosphere
40. field
41. Cretaceous-Tertiary

Chapter 7
The Moon

Answers to Problems in the Text

1. Use Fig. 7-2 plus a straightedge.
   a. noon b. sunset c. midnight d. sunrise
2. Full

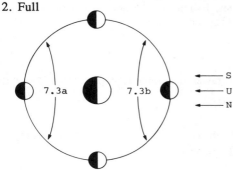

3. There are two possible answers since we could be in the Northern or Southern Hemisphere. If we are in the Northern Hemisphere, then picture 7.3a is correct, and we are seeing a third-quarter Earth. See Fig. 7-1.
4. The Moon rotates once about its axis for each revolution about the Earth. Because the Moon is not exactly spherical, the gravitational forces of the Earth have, over the eons, locked the Moon into synchronous rotation. These tidal forces slowed or sped up the Moon until it had synchronous rotation.
5. We would expect the tidal bulges to be under the Moon and on the far side of the Earth from the Moon (diametrically opposed). Because the Moon is overhead and in line with the Sun at noon when the Moon is new, the ideal tides would be maximum then and at midnight. But the real tides are unable to keep up with the Moon (they could if the oceans covered the world to a depth of 2–3 miles). In addition, the seacoasts cause resonance effects that also put the high tide out of phase with the Moon.
6. No. The orbit is inside the Roche limit, so we would expect the tidal forces of the Earth to separate them by overcoming their self-gravitation.
7. a. The inside edges of young craters, where no lava flow has occurred.
   b. Rougher because fresh rock surfaces are exposed. So many small impacts have occurred on older craters that the surface is covered with a smooth regolith.
   c. The crater Aristarchus.
8. a. No. The magnetic field is too weak for an ordinary compass.
   b. The Earth.
   c. It stays roughly fixed over the center of the front side, so it would appear to move very little. For an observer on the Moon, the angle between the observer's zenith and the Earth would be a measure of the distance of the observer from the point where the Earth is overhead.

9. The moon is less geologically active. No weathering or continental drift occurs on the Moon. No continental drifting is recirculating the surface material.

10. Sedimentary rocks are formed by air depositing dust or by running water depositing mud. These deposits turn into rock under pressure. For this to happen, there must be appreciable free water and an atmosphere.

11. The uplands are about 25% older than the maria. If the meteoritic cratering rate were constant, we would expect about 25% more craters on the uplands than on the maria and about the same distribution of sizes. Examination of photos shows that there are many more than 25% additional craters on the uplands.

Advanced Problems

12. a. Newton's law of gravity is that the gravitational force is proportional to Mm/r^2; that is,

$$F = G\,\frac{Mm}{r^2}.$$

Let M be the mass of a test body, $M_\oplus$ the mass of the Earth, and $r_\oplus$ the Earth's radius. Then $F_\oplus$, the weight on the Earth, is

$$F_\oplus = \frac{GM_\oplus m}{r_\oplus^2}.$$

If M is the mass of the Moon and $r_{\pmb{\jmath}}$ the Moon's radius, then the weight on the Moon, $F_{\pmb{\jmath}}$, is given by

$$F_{\pmb{\jmath}} = \frac{GM_{\pmb{\jmath}} m}{r_{\pmb{\jmath}}^2}.$$

If we divide one by the other, we get

$$\frac{F_{\pmb{\jmath}}}{F_\oplus} = \frac{GM_{\pmb{\jmath}} m}{r_{\pmb{\jmath}}^2} \times \frac{r_\oplus^2}{GM_\oplus m} = \frac{M_{\pmb{\jmath}}}{M_\oplus} \times \frac{r_\oplus^2}{r_{\pmb{\jmath}}^2}$$

$$= (0.012) \times \left(\frac{1}{0.27}\right)^2 = 0.165 = \frac{1}{6}.$$

b. Weight on Moon = $0.165 \times$ weight on Earth = $0.165 \times 180 = 29.7$ lbs (30 lbs if 1/6 is used).

13. Use the small angle equation $\dfrac{\alpha''}{2.1 \times 10^5} = \dfrac{d}{D}$, where α'' is the angular diameter of the Moon in seconds of arc, d is the lunar diameter, and D is the distance to the Moon.

$$\frac{\alpha''}{2.1 \times 10^5} = \frac{3476 \text{ km}}{384,000 \text{ km}} = 0.0091.$$

So $\alpha'' = 1.9 \times 10^3$ seconds of arc. To convert this to degrees, remember $60'' = 1'$ and $60' = 1°$, so $1° = 3600''$. $\alpha(°) = \dfrac{\alpha''}{3600} = 0°51$.

14. Escape velocity $= \sqrt{2}V_{circ}$; $V_{circ} = \sqrt{GM/R}$.
    Method 1

    $$V_{circ\text{☽}} = V_{circ\oplus}\sqrt{M_{\text{☽}}/M_\oplus} \sqrt{R_\oplus/R_{\text{☽}}} \text{ (from problem 12)}$$
    $$= 8 \text{ km/s} \times \sqrt{0.012} \times \sqrt{1/0.27}$$
    $$= 8 \text{ km/s} \times \sqrt{0.012/0.27}$$
    $$= 8 \text{ km/s} \times \sqrt{4.44 \times 10^{-2}}$$
    $$= (8 \times 0.21) \text{ km/s}$$
    $$= 1.68 \text{ km/s}$$

    So escape velocity $V_{esc} = \sqrt{2}V_{circ} = 2.38$ km/s.
    Method 2

    $$V_{circ\text{☽}} = \sqrt{GM_{\text{☽}}/R_{\text{☽}}} \qquad \begin{aligned} G &= 6.67 \times 10^{-11} \\ M &= 7.35 \times 10^{22} \text{ kg} \\ R &= 1.74 \times 10^6 \text{ m} \end{aligned}$$

    $$V_{circ} = \left(\frac{6.67 \times 7.35}{1.74} \times \frac{10^{-11} \times 10^{22}}{10^6} \right)^{1/2} \text{ m/s}$$
    $$= (2.86 \times 10^6)^{1/2} \text{ m/s} = 1.64 \times 10^3 \text{ m/s} = 1.64 \text{ km/s}$$

    Then escape velocity $= \sqrt{2} \times 1.64$ km/s $= 2.38$ km/s.

15. Mean density = mass/volume = $3M/(4\pi R^3)$. For Earth,
    $$M = 5.98 \times 10^{24} \text{ kg,}$$
    $$\text{radius} = \frac{\text{diameter}}{2} = 6.38 \times 10^6 \text{ m.}$$
    So mean density $= 5.5 \times 10^3$ kg/m^3.
    For the Moon,
    $$M = 7.35 \times 10^{22} \text{ kg,}$$
    $$\text{radius} = 1.74 \times 10^6 \text{ m.}$$
    So mean density $= 3.3 \times 10^3$ kg/m^3.
    Since iron is more dense than either of these numbers indicates, this implies the Moon has less iron than the Earth.

16. Use small angle
    $$\frac{\alpha''}{206{,}265} = d/D = \frac{\text{diameter of crater}}{\text{distance to Moon}}$$
    so
    $$d = \frac{D \times 2}{206{,}265} \text{ ,}$$
    where D is the distance to the Moon from Earth's surface; distance to Moon $= 3.48 \times 10^5$ km $= 3.48 \times 10^8$ m (less 0.06×10^8 m if you consider that you are on the surface of the Earth), so
    $$d = \frac{2 \times 3.48 \times 10^8}{2.06 \times 10^5} = 3.38 \times 10^3 \text{ m} = 3.38 \text{ km}$$
    (1.66 km at best by subtracting Earth's radius).

Sample Test Questions

True -False

1. The visible surface of the Moon has changed dramatically over the past million years. F
2. Because the same side of the Moon is always visible from the Earth, it is obvious that the Moon does not rotate. F
3. A full Moon rises at noon. F
4. The far side of the Moon is always dark. F
5. The sidereal period of the Moon is about 27-1/3 days. T
6. Most of the craters on the Moon were formed by recent impact. F
7. The tidal effects of the Earth on the Moon are causing the Moon to move away from the Earth. T
8. The Moon's tides are causing the Earth to rotate faster. F
9. The Roche limit is the distance from a planet within which a natural, large satellite will be disrupted by tidal forces. T
10. The smoother, gray, flat areas of the Moon are called maria. T
11. The lunar craters were formed by volcanoes. F
12. The first spacecraft reached the Moon in 1957. F
13. The lunar regolith is made primarily of debris blasted out of lunar craters as the craters formed. T
14. The Apollo 14 mission in 1971 landed near Fra Mauro. T
15. The Apollo 17 mission landed on the Moon in 1968. F
16. The oldest rock sample from the Moon has been dated at 4.5 billion years old. T
17. The bulk density of the Moon is less than that of the Earth. T
18. The spaceship landings on the Moon have shown that the Moon definitely has no iron-rich core. F
19. Most astronomers now believe the Moon was formed by fission from the Earth. F

Multiple Choice

Consider the diagram:

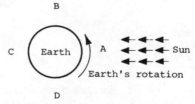

1. When the Moon is full it is in position ___C___.

2. When the Moon is new it is in position ___A___.

3. When the Moon is in first quarter it is in position ___B___.

4. When the Moon is in third quarter it is in position ___D___.

56

5. Because the Moon rotates on its axis with a period equal to its orbital revolution about the Earth, it is said to be in _____ rotation.
   - A. harmonic
   - B. synodic
   - *C. synchronous
   - D. elliptical
   - E. sporadic

6. The interval from new moon to new moon is not the sidereal period of the Moon because the
   - A. Moon is in synchronous rotation
   - *B. Earth goes about the Sun
   - C. Earth goes about the Moon
   - D. Earth produces tides on the Moon
   - E. mean density of the Moon is less than that of the Earth

7. The sidereal period of the Moon is about _____ days.
   - A. 13
   - *B. $27\frac{1}{3}$
   - C. $29\frac{1}{2}$
   - D. 37
   - E. 365

8. The period of lunar phases is about _____ days.
   - A. 13
   - B. $27\frac{1}{3}$
   - *C. $29\frac{1}{2}$
   - D. 37
   - E 365

9. Suppose you lived on the rim of the crater Copernicus on the Moon. The Earth would set below your horizon
   - A. once in 12 hours
   - B. once in 24 hours
   - C. every $27\frac{1}{2}$ days
   - D. every $29\frac{1}{2}$ days
   - *E. never

10. Ideally, the ocean tides should be only a bulge of water directly beneath the Moon.
    - A. correct
    - B. wrong: bulge at 90° to Moon position
    - *C. wrong: also a bulge diametrically opposed to the Moon
    - D. wrong: only a bulge diametrically opposed to the Moon
    - E. misleading: Ideally, there would be no ocean tides. The Earth's rapid rotation produces the tides.

11. The Moon keeps the same side of its surface toward the Earth as the Moon revolves about the Earth. This behavior is caused by the
    A. tides produced by the Moon
   *B. tides produced by the Earth
    C. tides produced by the Sun
    D. sporadic impact of meteorites on the Moon
    E. none of these

12. Tidal effects are causing the Moon to _____ the Earth.
    A. approach
    B. recede from
    C. speed up the rotation of
    D. slow down the rotation of
   *E. both B and D

13. The Roche limit is
    A. the closest distance that an artificial satellite can approach the Earth without being burned up by the atmosphere
    B. the maximum size a planet can be before its own tidal forces would destroy it
   *C. The shortest distance a natural satellite of any size can be from a planet without being destroyed by tidal forces
    D. the minimal precessional period a planet can have
    E. none of these

14. The Roche limit for the Earth is about _____ km.
    A. 6000   D. 384,000
    B. 8000   E. 971,000
   *C. 18,000

15. The Moon has a heavy atmosphere and much free water.
    A. correct
    B. wrong: little or no atmosphere
    C. wrong: little or no free water
   *D. both B and C
    E. Astronomers cannot yet answer this question.

16. The surface of the Moon is very cold even in the daylight.
    A. correct
    B. wrong: It is always hot.
    C. wrong: It is cold in daylight, hot at night or in the shade.
   *D. wrong: It is hot in the daylight, cold in the shade at night.
    E. We do not know.

17. The circular depressions on the lunar surface are called
    A. maria   D. rays
   *B. craters   E. mountains
    C. rilles

58

18. The lunar maria are
    A. cracks on the lunar surface
    *B. flat areas on the lunar surface
    C. bright streaks on the lunar surface radiating from the craters
    D. mountains on the Moon
    E. bodies of water on the Moon

19. The lunar craters were apparently caused by
    A. bursting bubbles of gas from the interior
    B. spacecraft landings
    *C. meteoritic collisions
    D. Nothing. There are in fact no craters on the Moon.
    E. Scientists have no good ideas of their cause.

20. The first Apollo mission to land on the Moon was Apollo _____.
    A. 1   *D. 11
    B. 5   E. 17
    C. 8

21. The Apollo 16 mission landed on the Moon near
    *A. the crater Descartes   D. Mare Imbrium
    B. Fra Mauro   E. the Taurus Mountains
    C. Mare Tranquillitatis

22. The first Apollo to use the lunar roving vehicle was Apollo _____.
    A. 8   *D. 15
    B. 11   E. 17
    C. 13

23. Most lunar rocks are
    *A. igneous   D. volatile
    B. sedimentary   E. none of these
    C. tidal

24. The mean density of an object is defined as
    A. total mass times total volume
    *B. total mass divided by total volume
    C. total volume divided by total mass
    D. total mass divided by total surface area

25. The _____ elements are those that would evaporate if the rocks containing
    them were strongly heated.
    A. sedimentary   *D. volatile
    B. igneous   E. basaltic
    C. refractory

26. Studying the Moon's surface show us that the cratering rate has been roughly constant for the last 4.6 billion years.
   A. correct
   *B. wrong: There was early intense bombardment.
   C. wrong: There has been recent intense bombardment.
   D. wrong: There was a break in bombardment 500 million years ago.
   E. misleading: We can learn nothing about the cratering rate by studying the present surface.

27. The layer of rocky soil covering much of the uplands of the Moon is sometimes called a _____. It was produced by _____.
   A. tsunamis/neap tides
   B. sediment/erosion
   C. lithosphere/meteoritic cratering
   D. megaregolith/volcanic action
   *E. megaregolith/meteoritic cratering

28. Which of the following is the new, post-Apollo theory of the Moon's origin?
   A. fission theory
   B. captive theory
   C. co-accretion theory
   D. sister-planet theory
   *E. impact-trigger theory

Mathematical Multiple Choice

*29. Given that the Moon is 3.48×10^5 km from the Earth, assume that you are on a spaceship halfway to the Moon and are using a telescope that enables you to resolve 0.1" of arc. What is the size of the smallest crater you can resolve?
   *A. 82 m
   B. 17 km
   C. 1.7 km
   D. 32 km
   E. 3.4 km

*30. If the command module of an Apollo mission is $1\frac{1}{2}$ times the lunar radius from the surface, how much must the speed be increased to reach escape velocity?
   A. 0.5 km/s
   *B. 0.8 km/s
   C. 2.38 km/s
   D. 8 km/s
   E. 1.94 km/s

*31. If a ball has a mass of 20 kg and a radius of 10 cm, then its density is about _____ kg/m^3.
   A. 2000
   B. 4.8
   C. 480
   *D. 4800
   E. 4,800,000

(Note: Students may miss the *cm* and take it as *m*.)

Essay

1. Describe how the Moon produces tides on the Earth. Use a diagram to clarify your explanation. (Expect figure similar to Fig. 7-4 and discussion similar to that on pp. 144–145.)
2. Describe how the Moon's orbit is affected by Earth tides and what will happen to the Earth's rotation. Describe how the Sun will affect the Earth-Moon system. (See pp. 145–146. Some main points: Moon moves out and Earth's rotation slows down. When the Moon is far enough away, the Sun slows the Earth enough so that the Earth comes into synchronous rotation with the Moon. Finally, the Sun slows the Earth even further so that the Moon moves back in closer to the Earth.)
3. On a photograph of the Moon (similar to Fig. 7-9), point out the terminator, limb, an impact crater, mare, terra, basin, ray, and rille. (See an identified photo or drawing.)
4. List the four ways of analyzing a rock. (See p. 154.)
5. Briefly describe the three old (pre-Apollo) theories of the Moon's origin. (See pp. 160–162: Fission—Moon broke off from Earth. Co-accretion—Moon formed in orbit around the Earth at the same time. Capture—Moon originated elsewhere in the solar system and was captured.)
6. Briefly describe the new (post-Apollo) theory of the Moon's origin. (See p. 163: Impact-trigger theory—after Earth's core separated, a large meteorite blasted off particles of the Earth's crust, which formed the Moon.)

Word Practice

1. _____ _____ is mass divided by volume.

2. The _____ _____ _____ pulverized the Moon's surface.

3. The _____ theory states that the Moon formed from particles produced from an impact that occurred after the Earth formed.

4. The _____ lunar formation theory considers that the Moon formed in orbit around the Earth as the Earth formed.

5. The _____ is the line separating the light and dark sides of the Moon.

6. The _____ were mostly formed on the Moon by impacts.

7. _____ _____ would evaporate if the rocks they were in were strongly heated.

8. The _____ _____ of the Moon is a result of tidal forces.

9. _____ of interplanetary debris formed craters.

*10. The _____ is that part of the Moon beneath the crust.

11. _____ are large sea waves produced by earthquakes at sea.

*12. The _____ Moon rises at midnight.

13. The _____ _____ is the lunar rotation about the Earth with respect to the stars.

14. _____ _____ have high boiling temperatures.

15. _____ _____ is movement of the Moon away from the Earth caused by the Earth's relatively rapid rotation.

16. The _____ are bright streaks radiating from craters.

17. The _____ _____ is a bulge in the ocean on the front and back sides of the Earth.

18. The _____ theory considers that the Moon originated elsewhere in the solar system.

19. _____ _____ is the minimum distance between two bodies such that the larger is able to tidally disrupt the smaller.

20. The lunar origin theory in which the Moon broke off from the Earth is the _____ theory.

*21. The _____ Moon rises at sunset.

22. The lunar origin theory in which the Moon and Earth formed nearby and at the same time is the _____ theory.

23. The stretching of the Earth produced by the Moon is called _____ _____.

24. A(n) _____ is a smooth, gray patch on the Moon's surface.

25. _____ are distortions in shape caused by gravitational forces.

26. The edge of the Moon as seen from Earth is the _____.

27. _____ are vast craters of the Moon.

28. _____ are rocks composed of cemented rock chips.

29. A body that rotates about a larger body is often called a(n) _____ of the larger body.

30. The different shapes that the Moon can appear to have are called the _____ of the Moon.

31. The _____ _____ Moon is invisible at any time.

32. The _____ Moon occurs seven days after the new Moon.

33. A(n) _____ occurs between a quarter and a full Moon.

1. mean density
2. early intense bombardment
3. impact-trigger
4. co-accretion
5. terminator
6. craters
7. volatile elements
8. synchronous rotation
9. impact
10. lithosphere
11. tsunamis
12. third-quarter
13. sidereal period
14. refractory elements
15. tidal recession
16. rays
17. ocean tide
18. capture
19. Roche's limit
20. fission
21. full
22. co-accretion
23. body tide
24. mare
25. tides
26. limb
27. basins
28. breccias
29. satellite
30. phases
31. new Moon
32. first-quarter
33. gibbous

Chapter 8
Introducing the Planets—Mercury

Answers to Problems in the Text

1. Venus
2. 11.2; 318
3. See Table 8-1, satellite Titan of Saturn
4. Charon of Pluto—next is the moon of our Earth
5. Pluto, but only because of its large eccentricity. On the average the Earth comes closer.

6.

Property	Terrestrial Planets	Gas Giant Planets
Size	small	large
Temperature	warm	cold
Density	moderate	low
Number of satellites	few or none	many
Orbital properties	near Sun	far from Sun
Composition (surface)	silicates, metals	ices, gas
Atmosphere	hydrogen-poor	hydrogen-rich
Surface	solid,rocky	icy, liquids

7. *Common to Mercury and the Moon:* cratering, pulverization of the surface layers, and lava flows. *Unique to Mercury:* a definite core and a magnetic field generated by currents flowing in the core, no indication of present or past plate tectonic activity. The lack of tectonic activity on Mercury indicates that the heat flow and and mantle convection must be smaller than the Earth's.
8. These two planets are above the horizon only during the daytime or near the horizon after sunset or before sunrise. When the planets are near the horizon, the Earth's atmosphere makes observation difficult. Thus, daytime observation may be best. At midnight they are both below the horizon.
9. Both the Moon and Mercury are low-pressure environments, so the suits will need to be airtight. The temperature extremes on Mercury are much greater and the ultraviolet radiation is much more intense, so the Apollo suits would need extra shielding for radiation and against thermal losses.

10. Assume orbital velocities are circular velocities:

$$V_{circ} = \sqrt{GM/R}.$$

For planet orbiting Sun, use solar mass $= 2.0 \times 10^{30}$ kg, $G = 6.67 \times 10^{-11}$ SI units.

$$\begin{aligned} V_{circ} &= (6.67 \times 10^{-11} \times 2.0 \times 10^{30})^{1/2}/R^{1/2} \\ &= 1.19 \times 10^{10}/\sqrt{R}, \end{aligned}$$

so
$$\begin{aligned} V_{\oplus} &= 1.19 \times 10^{10}/\sqrt{1.5 \times 10^{11}} = 3.08 \times 10^4 \text{ m/s} = 31 \text{ km/s} \\ V_p &= 1.19 \times 10^{10}/\sqrt{5.9 \times 10^{12}} = 4.9 \times 10^3 \text{ m/s} \\ &= 4.9 \text{ km/s.} \end{aligned}$$

11. Use $F = GM_{\oplus}M_M / R_{\oplus}^2$, where $R_{\oplus} = 1$ AU = Earth-Sun distance; replace $M_{\oplus}$ by $4M_{\oplus}$ and $R_{\oplus}$ by $2R_{\oplus}$—no change in result.

12. a. We want d in $\alpha''/206,265 = d/D$ if $\alpha'' = 1''$ and $D = 135 \times 10^6$ km.

$$\begin{aligned} d = (\alpha''D)/206,265 &= (1 \times 1.35 \times 10^{11})/(2.06 \times 10^5) \\ &= 6.55 \times 10^5 \text{m.} \end{aligned}$$

We can resolve 655 km.

b. The side toward us is unlit; that is, Mercury is in new phase. Also, we would be looking almost directly at the Sun.

c. Assume what is wanted is d in $\alpha''/206,265 = d/D$ with $\alpha'' = 1''$ and $D = 0.277$ AU $= 4.2 \times 10^{10}$ m. Then

$$d = (1 \times 4.2 \times 10^{10})/(2.06 \times 10^5) = 2.04 \times 10^5 \text{ m.}$$

A thickness of 209 km could be resolved.

13. The relevant densities are 5500 kg/m^3 for the Earth and 3300 kg/m^3 for the Moon. Either look these up or use formula below. Use Table 8-1:

density of Mercury $= 3M /4\pi r^3 = 5400$ kg/m^2,

which is closer to Earth's value than the Moon's, but Mercury is closer to the Moon in size.

14. a. Mass of Mercury = density of core × volume of core + density of mantle × [volume of Mercury — volume of core]

but mass of Mercury – mean density × volume of Mercury.

Equate and divide by volume of Mercury, giving

$$\text{mean density} = \text{core density} \times \left[\frac{\text{volume of core}}{\text{volume of Mercury}}\right] + \text{mantle density}$$
$$\times \left[1 - \frac{\text{volume of core}}{\text{volume of Mercury}}\right].$$

b. Let $\bar{p}$ be mean density, ρ_c core density, ρ_m mantle density, R_c core radius, and R_m Mercury's density.

$$\frac{\text{volume core}}{\text{volume of Mercury}} = \frac{\frac{4}{3}\pi R_c^3}{\frac{4}{3}\pi R_m^3} = \left(\frac{R_c}{R_m}\right)^3$$

and

$$\overline{\rho} = \rho_c \left[\frac{R_c}{R_m}\right]^3 + \rho_m \left[1 - \left(\frac{R_c}{R_m}\right)^3\right]$$

Solving for R_c,

$$R_c = R_m \sqrt[3]{\frac{\overline{\rho} - \rho_m}{\rho_c - \rho_m}} = R_m \sqrt[3]{\frac{5500 - 3300}{8000 - 3300}} = 0.78 R_m.$$

The radius of the core of Mercury is 0.78 times the total radius, a much larger fraction than for the Earth.

Sample Test Questions

True-False

1. The symbol for the Sun is ⊙. T
2. The symbol for Saturn is ⊕. F
3. The four inner planets are called terrestrial planets. T
4. The diameter of Jupiter is about 10 times the diameter of the Earth. T
5. The terrestrial planets have more methane (CH_4) than the giant planets. F
6. Superior planets are closer to the Sun than the Earth is. F
7. Only superior planets can be in opposition. T
8. A planet in opposition rises at sunset. T
9. The Ptolemaic model had the Sun at the center of the universe. F
10. Jupiter is slightly more than five times as far away from the Sun as the Earth is. T
11. The orbit of Mercury is closer to lying in the ecliptic plane than is that of any other planet. F
12. The orbital precession of Mercury is about 43 seconds of arc per century. F
13. The orbital precession of Mercury beyond that predicted by Newton's laws is about 43 seconds of arc per century. T
14. The daytime temperature of the surface of Mercury is above 400 K. T
15. The temperatures on the surface of Mercury are so high in the daytime that scientists expect there to be pools of molten metal. F

Multiple Choice

1. The terrestrial planets are, in order of increasing distance from the Sun:
   A. Mars, Venus, Earth, Mercury
   *B. Mercury, Venus, Earth, Mars
   C. Mercury, Mars, Venus, Earth
   D. Earth, Mercury, Mars, Venus
   E. Venus, Mercury, Earth, Mars

2. Which planet on this list has no satellites?
   A. Mars   *D. Venus
   B. Earth   E. Uranus
   C. Jupiter

3. The radius of Saturn is about _____ times the Earth's radius.
   A. 1   *D. 9
   B. 10   E. 4
   C. 318

4. Venus is about _____ AUs from the Sun.
   A. 10   *D. $\frac{3}{4}$

   B. 1.5   E. $\frac{1}{2}$

   C. 1

5. Which of the following is a satellite of a planet?
   A. Venus   *D. Phobos
   B. Uranus   E. none of the above
   C. Ceres

6. As Mercury revolves about the Sun, it
   A. keeps the same side toward the Sun
   B. rotates retrograde
   *C. keeps similar sides to the Earth and Sun during every third apparition
   D. emits radar waves
   E. does not rotate

7. The day on Mercury is about _____ Earth days.
   A. 1   D. 159
   *B. 59   E. Mercury always keeps the same side
   C. 88   toward the Sun.

8. The first spacecraft to pass near Mercury was _____ in 1974.
   A. Apollo II   *D. Mariner 10
   B. Venera 9   E. Sun Probe 7A
   C. Mercury V

9. The orbit of Mercury was puzzling because _____. This effect was
   explained by general relativity.
   A. it was highly inclined
   B. it was the least circular of the planetary orbits
   C. it regressed
   *D. the orbital precession was larger than Newton's laws predicted
   E. the orbital precession was smaller than Newton's laws predicted

10. The larger the planet, the _____ its surface features are likely to be.
    A. larger   D. older
    B. smaller   *E. younger
    C. redder

11. The _____ the planet, the more likely there will be internal geological activity.
    A. faster
    B. slower
    C. smaller
    *D. larger
    E. closer to the Sun

12. The day on Mercury is _____ Earth days.
    A. 5
    *B. 59
    C. 88
    D. 100
    E. Mercury always keeps the same side toward the Sun.

Essay

1. Using photographs of Mercury and the Moon, circle similarities and note differences. (Instructor: Use photographs similar to those in Chapter 8 for Mercury and Chapter 7 for the Moon.)
2. Describe the difference between terrestrial and giant planets. (Instructor: See Problem 8-6.)
3. Discuss the idea of comparative planetology. (See pp. 171–172.)
4. Discuss how surface features and planet size are related. (See pp. 181–182.)

Word Practice

1. The closest planet to the Sun is _____.

2. _____ is the next planet farther away from the Sun than the Earth.

3. _____ has an orbit between the Earth and Jupiter.

4. The third planet out from Jupiter is _____.

5. The planet _____ has rings easily visible with a small telescope.

6. The four large outer planets of the solar system are often called the

   _____ _____.

7. _____ is the second planet from the Sun.

8. _____ is the seventh planet from the Sun.

9. _____ are solid bodies orbiting around planets.

10. _____ is the largest planet in the solar system.

11. The farthest known planet from the Sun on the average is _____.

12. Our Sun is a(n) _____.

13. _____ _____ is the comparison of different planets to try to

    understand why the planets are similar or different.

14. The small shift in Mercury's perihelion position left after classical effects are removed is a non-Newtonian _____ _____.

15. The study of the origins and development of planets is called _____.

16. About 4 billion years ago there was a(n) _____ _____ _____ that left the planets heavily cratered.

Answers to Word Practice

1. Mercury
2. Mars
3. Mars
4. Neptune or Pluto
5. Saturn
6. giant planets or gas giants
7. Venus
8. Uranus

9. satellites
10. Jupiter
11. Pluto
12. star
13. comparative planetology
14. orbital precession
15. planetology
16. intense early bombardment

Chapter 9
Venus

Answers to Problems in the Text

1. Although Mercury is closer to the Sun, Venus is hotter because of the greenhouse effect (trapped infrared rays).

2. Early in its history, Venus probably reached a temperature high enough that most of the H_2O was broken up into H_2 and O_2. The H_2 was lost and much of the oxygen absorbed by the hot surface (oxidation), so H_2O could not reform. On Earth the temperature was low enough that the H_2O remained. When the Earth cooled, the H_2O condensed into a liquid, removing most of it from the atmosphere. On Earth the H_2O could dissolve the CO_2; without water, Venus kept its primordial CO_2 in its atmosphere.

3. A spacesuit on Venus must keep the atmosphere out and keep the pressure from crushing the wearer. Present astronaut suits would not work on Venus. We would need a rigid suit to withstand the atmospheric pressure.

4. a. On Venus, the flow patterns are roughly parallel to the rotation axis, while on Earth they are more curled.
   b. How the wind patterns are affected by the rotation of the planet. Also, effects of density.
   c. Coriolis force is smaller on Venus. We would expect less spiraling of the wind patterns.

5. Without intervening clouds of water vapor, most of the Sun's radiation can reach the surface of the desert in the daytime, causing the temperature to be high. At night the heat can radiate away more rapidly because there is no water vapor to absorb and return (by reradiation) the infrared radiation that leaves the surface. In other climates where the air is moist or cloudy, the clouds absorb the Earth's nighttime infrared radiation, thus keeping the air from cooling so rapidly.

6. Primitive features are (a) and (c); evolved features are (b), (d), and (e). In order of increasing geologic evolution: Moon, Mercury, Venus, Earth.

7. In general, the more massive a planetary body, the more rapidly its surface geologically evolves.

8. There is little chance of life on Venus. There is nothing of any density to serve as a solvent medium (as water does on Earth). Also, the temperature is too high for life as we know it.
   a. If these conditions existed, chances would be good. (See discussion of the origin of life in Chapter 28.)
   b. No, but current beliefs about evolution make the existence of life under earthlike conditions very likely.

9. Closest approach: 0.277 AU $= 4.2 \times 10^{10}$ m. Moon's distance: 3.48×10^8 m. Venus is about 110 times as far away as the Moon.

10. Venus must be near inferior conjunction; that is, it appears very close to the Sun. So Venus is almost between the Earth and Sun. It cannot be near superior conjunction because then its phase would be nearly full.

11. Use small angle formula with $\alpha" = \frac{1}{2}$; that is, $\frac{\alpha"}{2.1 \times 10^5} = \frac{d}{D}$ where d is cloud thickness and D is distance. From Table 8-1, distance = 1 AU–.723 AU = 0.28 AU = $0.28 \times 1.5 \times 10^{11}$ m = 4.2×10^{10} m.

So $d = \dfrac{\alpha"D}{2.1 \times 10^5} = \dfrac{1}{2} \times 4.2 \times 10^{10} \dfrac{m}{2.1 \times 10^5} = 1.0 \times 10^5$ m.

Clouds must be more than 100,000 m = 100 km thick.

12. Use Wien's law.
$$W = \frac{0.00290}{T}, T = 750 \text{ K},$$
so $W = (2.90 \times 10^{-3})/(7.5 \times 10^2 \text{ K}) = 3.9 \times 10^{-6}$ m.
This wavelength corresponds to infrared radiation.

13. Mean density $= \dfrac{\text{mass}}{\text{volume}} = \dfrac{\text{mass}}{\frac{4}{3}\pi r^3} = \dfrac{6 \times \text{mass}}{\pi d^3}$. (Get mass and radius from Table

8-1.) Density $= 4.87 \times \dfrac{10^{24} \text{ kg}}{(1.2 \times 10^7 \text{ m})^3} \times \dfrac{6}{\pi} = 5.4 \times 10^3$ kg/m^3. Since this is slightly lower than the Earth's mean density, we might expect less than the Earth's amount of iron.

Sample Test Questions

True-False

1. A planet with a prograde rotation rotates east to west. F
2. Venus has a retrograde rotation. T
3. The atmosphere of Venus is mostly nitrogen. F
4. The surface of Venus is hot and wet. F
5. Venus has a surface atmospheric pressure about 90 times that of the Earth's atmosphere. T
6. The clouds of Venus are water clouds like the Earth's. F
7. The surface of Venus is heated by the greenhouse effect. T
8. The Urey reaction explains why there is no free water on the surface of Venus. F
9. The Urey reaction explains the low level of CO_2 in the Earth's atmosphere. T
10. The study of the origins and development of the planets is termed planetology. T
11. The smaller the planet the younger its surface features are likely to be. F

1. Venus rotates east to west (retrograde). This was discovered by radar.
   - *A. correct
   - B. wrong: Rotation must be from west to east to be called retrograde.
   - C. wrong: It was discovered by optical telescopes.
   - D. wrong: It was discovered by a Soviet probe landing on Venus.
   - E. wrong: Both B and D corrections are needed.

2. The atmosphere of Venus is mostly
   - A. oxygen
   - B. nitrogen
   - *C. carbon dioxide
   - D. neon
   - E. sulfuric acid

3. The clouds of Venus are composed of droplets of
   - A. water
   - B. nitrogen
   - C. carbon dioxide
   - D. salt
   - *E. sulfuric acid

4. A cool, large planet can most easily keep its atmosphere.
   - *A. correct
   - B. wrong: The planet must be cool and small to keep its atmosphere.
   - C. wrong: The planet must be hot and large to keep its atmosphere.
   - D. wrong: The planet must be hot and small to keep its atmosphere.

5. For a given temperature, the lighter the gas molecules, the _____ their average speeds.
   - A. slower
   - *B. higher
   - C. bluer
   - D. more primitive
   - E. misleading: Temperature does not affect speed.

6. Venus is said to be hot because of a greenhouse effect. This means
   - A. Venus is covered with glass
   - B. Venus is green
   - C. radiation can't get to the surface of Venus since it has clouds
   - *D. the carbon-dioxide clouds keep the surface heat of Venus from radiating
   - E. both A and B

Essay

1. Discuss conditions on the surface of Venus. (See pp. 188–190.)
2. What is the difference between rotation and revolution? (See Fig. 9-1.)
3. What three principles govern the motions of gas molecules in atmospheres? (See p. 194.)
4. Discuss the atmosphere of Venus: composition, clouds, density, pressure. (See p. 185.)

Word Practice

1. The _____ _____ _____ _____ means that Venus rotates east to west.

2. The study of the origins and development of planets is termed _____.

3. The atmosphere that a planet develops as it outgasses and evolves is called its

   _____ _____.

4. The heating of a planet's atmosphere caused by CO_2 trapping infrared rays is called

   the _____ _____.

5. The total energy striking a planet per unit area per second is the planet's

   _____ _____.

6. The initial gaseous atmosphere of a planet is the planet's _____

   _____.

7. A planet that rotates from west to east is said to have _____ rotation.

8. The clouds of Venus are mostly droplets of _____ _____.

Answers to Word Practice

1. retrograde rotation of Venus
2. planetology
3. secondary atmosphere
4. greenhouse effect
5. solar constant
6. primitive atmosphere
7. prograde
8. sulfuric acid

Answers to Problems in the Text

1. Main points: Day length ~ 24 h. Landscape: reddish soil with many rock fragments. Either still (19 mph) or strong winds (greater than 100 mph). If no clouds, sky reddish, occasional dust storms. Color of sky depends upon amount of dust. Temperature changes of 60° C from sunrise to midafternoon. May get a few degrees above freezing on a summer afternoon. Remember, Vikings landed at "dullest" or flattest place. There are impressive cliffs, canyons, craters, snowfields, and ice caps elsewhere. Might see one or two small moons in the sky.

2. Mars is more earthlike than the Moon. With frozen water and past free water, there is a chance of finding life, evidence of past life, or evidence of conditions that prevented life from arising. A colony on Mars could obtain its water by heating Martian soil, whereas the Moon apparently contains no water.

3. Craters on Mars show signs of erosion (by water for the oldest craters and by dust for the younger ones). In addition, on Mars we see the dark dust deposits downwind from the craters and in some crater interiors.
   a. Students should point out that on Mars erosion has caused "soft"-looking shallow craters with low or broken rims.
   b. The Martian environment has caused more topological changes—for example, erosion.

4.

	Mercury	*Mars*	*Earth*
Atmosphere	little or none	low pressure	1 atmosphere
Craters	surface well covered by craters	older craters are much eroded	most craters are eroded away
Plate tectonics	none	possible: break in one large canyon	active tectonics
Volcanism	volcanic lava flows form mare plains but no volcanic mountains	large, isolated, young volcanoes and lava flows covered by dust	active volcanic mountains; volcanoes in chains
Surface	old	mixed	young

5. These satellites can be used to answer questions such as:
   Are the two moons of the same age and composition? Are they pieces of a single large primeval moon? Are they of different origins? Are they perhaps captured asteroids? Do they clarify the age and origin of our moon? Do their surfaces reflect the past history of the solar system?

6. a. The sample would most likely be seabed. Most seabed is sedimentary with organic sediment. About one out of six would land on the solid surface. They would probably get widely different soil types. However, the two Vikings found virtually identical soil at two widely separated sites, which implies that the dust is well mixed by the wind.

   b. The Earth is divisible into several distinct types off terrain—arctic, plains, mountains, desert, tropic, and oceans—so 15–24 well-placed landings (three in each type of terrain) would give only a gross idea. If landings were random, 90–144 would be needed because most would be lost at sea. If we wanted to characterize people or other life forms, more would be needed.

   c. If landings were random, fewer might be needed—possibly 20–40, well-placed, should give us a good characterization. Mars is smaller and it does not appear to be as different in various regions, yet there are volcanoes, canyons, sand dunes, rocky plains, dark and light soil, two kinds of ice cap, and so on, which we would like to understand.

Advanced Problems

7. Use weight = force = $GMm/r^2 = W$, then
   $$W_{Phobos}/W_{Earth} = (M_{Phobos}/M_{Earth})/(r_{Phobos}/r_{Earth})^2 .$$
   Use Table 8.1—since Phobos is not spherical, use an average value, say 12 km. The ratio then becomes
   $$W_{Phobos}/W_{Earth} = 4.5 \times 10^{-4};$$
   hence, $W_{Phobos} = 0.059$ lbs = 0.9 oz.

8. $t = \sqrt{2h/a}$; $a = 0.0042$ m/s
   a. Assume eye level is 1.70 m from ground (6 feet to top of head).
   $$t = \sqrt{2 \times 1.70/0.0042} \text{ s} = \sqrt{810} \text{ s} \approx 28 \text{ s}$$
   b. $t = \sqrt{2 \times 1.70/10}$ s $= \sqrt{34 \times 10^{-4}}$ s = 0.058 s

9. $\dfrac{\alpha"}{206,265} = d/D$; $\alpha" = 1"$;
   $D = 56 \times 10^6$ km = 5.6×10^{10} m;
   $$d = (\alpha"D)/(2.06 \times 10^5) = \frac{5.6}{2.06} \times 10^5 \text{ m} = 2.7 \times 10^5 \text{ m}$$
   Objects larger than 270 km could be resolved. Lowell did not claim to see canals but rather what he thought was vegetation along the sides.

10. $V_{circ} = \sqrt{GM/R}$
    $G = 6.67 \times 10^{-11}$ SI; $M = 6 \times 10^{15}$ kg;
    $R = 8$ km $= 8 \times 10^3$ m
    $V_{circ} = [2(6.67 \times 6/8) \times 10^{-11} \times 10^{15}/10^3)]^{1/2} = 7$ m/s
    If an object were not too massive, it could easily be put in circular orbit by hand.
    $V_{escape} = \sqrt{2}V_{circ} = 10$ m/s, so the object could also easily escape.

Sample Test Questions

True-False

1. The first successful Mars landing was Viking 1 in 1976. T
2. Mars appears as a uniformly reddish ball through a telescope. F
3. Mars shows seasonal changes. T
4. The term *canali* was first popularized by Lowell. F
5. The reddish color of Mars is caused by iron oxide. T
6. The sky of Mars is reddish-tan. T
7. Daytime air temperatures at the two Viking sites got above freezing (273 K). T
8. The air pressure at the surface of Mars is about 0.7% of that on Earth. T
9. The Martian polar caps consist of a permanent cap of CO_2 with an overlayer of water ice that melts each year. F
10. The changing dark markings on Mars are not yet understood. F
11. The volcanoes of Mars are more numerous and smaller than those on the Earth. F
12. There is no evidence that there has ever been free water on Mars. F
13. The present lack of free water on Mars may have been caused by the volcanoes around Olympus Mons. T
14. The Viking experiments that were designed to directly detect organic molecules in the Martian soil did not detect any. F
15. Phobos is the inner satellite of Mars. T
16. Phobos is definitely an asteroid captured by mars. F

Multiple Choice

1. The rotation period of Mars is about
   - A. 2 hours, 10 minutes
   - B. 6 hours
   - C. 3 days
   - D. 59 days
   - *E. not near any of these

2. The dark areas of Mars are called
   - A. Martian dark areas
   - B. deserts
   - *C. maria
   - D. permafrost
   - E. canals

3. The first spacecraft to reach the surface of Mars was
   - A. Viking 1
   - *B. Mars 2
   - C. Mars 1
   - D. Mariner 10
   - E. Pioneer 10

4. Both of the Vikings landed on rocky plains.
   - *A. correct
   - B. wrong: Only Viking 1; Viking 2 landed on a smooth, featureless plain.
   - C. wrong: Only Viking 2; Viking 1 landed on a smooth, featureless plain.
   - D. wrong: Both Vikings landed on smooth plains.
   - E. wrong: Only one Viking successfully landed.

5. The Martian air temperatures at the Viking landing sites ranged from a high of about 244 K to a low of 87 K.
   A. correct
   B. wrong: The high was 576 K.
   *C. wrong: The low was 187 K.
   D. wrong: Both B and C changes are needed to be correct.
   E. misleading: The Vikings were not equipped to measure air temperatures.

6. The Martian atmosphere is mostly CO_2. The next two compounds in order of amount are O_2 and argon.
   A. correct
   B. wrong: It is mostly H_2O.
   *C. wrong: The next two are nitrogen and argon.
   D. wrong: Both B and C changes need to be made.
   E. misleading: Only the major component, CO_2, is known.

7. The most common gas in the atmosphere of Mars is
   *A. CO_2 D. H_2O
   B. N_2 E. none of these
   C. O_2

8. The air pressure at the Viking sites was about ____% of the Earth's normal pressure.
   A. 700   *D. 0.7
   B. 70   E. 0.07
   C. 7

9. That Martian volcanoes are larger than those on Earth is due to
   A. the weaker surface gravity on Mars
   *B. the lack of plate tectonics on Mars
   C. the lowering of the lava-melting temperature by the weak Martian atmosphere
   D. the greater distance of Mars from the Sun
   E. misleading: Martian volcanoes are smaller, not larger.

10. The channels on Mars discovered by Mariner 9 show
    A. that intelligent life once existed on Mars
    B. that there were flaws in the Mariner 9 design
    *C. that there most likely once was free water on Mars
    D. That Schiaparelli was right
    E. misleading: There are no channels on Mars.

11. It is believed that beneath the surface of Mars lies a layer of frozen soil containing water ice that does not melt in the Martian summer. Such a layer is called a
    A. frozen layer
    B. maria
    C. channel
    *D. permafrost
    E. misleading: Most scientists do not believe this.

12. All of the Viking experiments could be interpreted to show that life exists, or at least once existed, on Mars.
   A. correct
   *B. wrong: Organic molecules were not found, and other experiments were ambiguous.
   C. wrong: All biological experiments were strongly negative.
   D. wrong: Only one experiment positively detected life.
   E. misleading: No Viking experiment checked this.

13. The Vikings and Mariners were able to show that Mars' moons Phobos and Deimos were once asteroids.
   A. correct
   B. wrong: They showed that they were once part of Mars.
   C. wrong: The moons are Phobos and Lowellos.
   D. wrong: Both B and C changes are needed.
   *E. misleading: The origin of the moons of Mars is not yet settled.

Problems Involving Optional Mathematical Equations from Other Chapters

*14. (Optional Equation III) Calculate the escape velocity from Deimos starting from a mountain 8 km from the center if its mass is 6×10^{15} kg.
   A. 1 km/s   D. 15 km/s
   B. 7 km/s   E. 30 km/s
   *C. 10 km/s

*15. (Optional Equation I) Mars at its closest is about 56 million km from Earth. Deimos is about 12 km in diameter and Phobos at largest is about 28 km long. Can either satellite be resolved (can their shape be seen) using an Earth-based telescope with a resolution of 1"?
   A. can resolve both
   B. can resolve Phobos but not Deimos
   C. can resolve Deimos but not Phobos
   *D. cannot resolve either

Essay

1. Describe the physical effect that caused the belief that Mars had canals. (See Fig. 10-4, p. 207.)
2. With photographs of Mars (Viking orbiter) and the Moon, compare the surfaces and note three major differences. (Main points: erosion of craters; channels; dust patterns in lees of craters.)
3. Explain how Mars could have lost its free water. (See pp. 213–216.)
4. Discuss the Martian surface features (craters, volcanoes). What do they tell us about the "geology" of Mars? (See pp. 209–212.)
5. Discuss evidence for or against life on Mars. (See pp. 216–218.)

Word Practice

1. There are _____ on Mars caused by meteoroid impacts.

2. The Viking spacecraft found no evidence for _____ _____
_____.

3. _____ _____ containing carbon are essential building blocks of
life as we know it.

4. A dark area on the surface of Mars is called _____.

5. The bright orange areas of Mars as seen through a telescope are called
_____.

6. The shrinking of the northern Martian polar ice cap in summer and the growth of
dark markings on the Martian surface are examples of
_____ _____.

*7. Mars was probably first seen through a telescope by _____.

8. The _____ is a possible layer of ice-rich soil a few meters beneath the
Martian surface.

*9. Mars' outer satellite is called _____.

10. The _____ of Mars are the largest known in the solar system. Olympus
Mons is the largest.

11. The darkening of the Martian surface is caused by _____ storms.

12. The Martian atmosphere is mostly _____ _____.

13. The observatory at Flagstaff, Arizona, was founded by _____ to study
Mars.

14. The first probe to reach Mars was the Russian probe _____
_____ , which crashed in 1971.

15. The first successful landings of spacecraft on Mars were the
_____ _____.

16. Both the surface and the sky of Mars appear _____.

17. The summer _____ _____ of Mars were measured between
187 K and 244 K by the Vikings.

18. The _____ of Mars is mostly CO_2 at low density.

19. The _____ _____ on Mars is about 0.7% of the Earth's.

20. The surface of Mars has _____ that look similar to dry river beds.

21. The inner satellite of Mars is named _____.

22. The _____ of Mars are an optical effect of poor observing conditions combined with the physiological tendency of the eye to connect dots into lines.

23. Mars's rotation _____ is $24^h 37^m$.

24. Rocks formed by melting within a planet are called _____ when they reach the surface.

25. The Martian _____ are rocks found on Earth that are believed to come from Mars.

Answers to Word Practice

1. craters
2. life on Mars
3. organic molecules
4. maria
5. deserts
6. seasonal changes
7. Galileo
8. permafrost
9. Deimos
10. volcanoes
11. dust
12. carbon dioxide
13. Lowell
14. Mars 2
15. Viking missions
16. reddish
17. air temperatures
18. atmosphere
19. air pressure
20. channels
21. Phobos
22. canals
23. period
24. lava
25. meteorites

Answers to Problems in the Text

1. a. Each of these planets has a different dimension, temperature, atmospheric pressure, rotation rate, and, most important, gravity, so they enable us to better understand how wind patterns depend upon these factors.

   b. Temperature, composition, atmospheric pressure, rotation period, planetary mass (gravity), and radius. The substances forming the "clouds" have different molecular weight and chemistry.

   c.

	Venus	Earth	Mars	Jupiter
Temperature	high	normal	low	low to moderate
Composition	CO_2	O_2, N_2	CO_2	H_2, He
Pressure	high	normal	low	high
Rotation period	slow	24^h	24^h	9^h
Planetary mass (Earth)	1	1	1/9	318
Surface features	some	some	high volcanoes	none known, if solid surface exists
Surface gravity/ Earth's	0.9	1	0.4	2.6

   d. Mercury has no atmosphere; Uranus, Neptune, and Pluto are extremely cold and hard to study, so no data are available about cloud patterns and the like.

2. a. Jupiter
   b. Pluto
   c. Venus or Jupiter, depending on your definition of "atmosphere"
   d. Jupiter—might argue Uranus
   e. Earth

3. Earth. Ancient and possible future environments on Mars. Possibly, the lower atmosphere of Jupiter if warm enough. Satellites of outer planets are icy; if local volcanic or other geothermal activity occurs there, water might temporarily appear. There may be enough tidal heating to provide a liquid layer on Europa under its icy crust.

4. Gravity is proportional to M/R^2, so ratio

   $$(G_{Io}/G_{Moon}) = (M_{Io}/M_{Moon}) (R_{Moon}^2/R_{Io}^2).$$

   Each Jovian satellite has about the same gravity as the Moon, with Io having a slightly stronger gravity and the others slightly weaker.

gravity	Io	Europa	Ganymede	Callisto
gravity of Moon	1.1	0.9	0.9	0.7

Europa, Ganymede, and Callisto have properties rather like a snow-spotted Moon. Io has some "exotic" properties, such as yellow-glowing sky.

5. Surface probably cratered; crunchy frozen ammonia frost and salt deposits underfoot; reddish and yellowish surface with possibly some stony soil in places. The sky would have a faint yellowish glow, its brightness depending on the position of Io in its orbit. Volcanic eruptions and sulfur flows.

6. a. The one farthest from the Sun because it would be cooler and hence the hydrogen would have less of a chance to escape.

  b. The one with more volcanism should have more CO_2, which is commonly released by volcanoes—at least in bodies with interior regions resembling those of terrestrial planets.

Advanced Problems

7. Distance to Jupiter at closest approach:
  $4 \text{ AU} = 4 \times 1.5 \times 10^{11} \text{ m}$

  a. Let us assume that real separation is $2 \times 10^6 \text{ km} = 2 \times 10^9$ for satellite-Jupiter distance. Then

  $\alpha''/206,265 = d/D = 2 \times 10^9 \text{ m}/(6 \times 10^{11}\text{m})$;

  $\alpha'' = \dfrac{2.06}{3} \times 10^{3''} = 687'' = 0\overset{\circ}{.}2$

  b. $0\overset{\circ}{.}2$ is about one-third the diameter of the full moon and well above the limit of resolution of the human eye, which therefore would easily see the satellites if they were bright enough. But Jupiter is so bright that its satellites are lost in the glare.

8. Take distance to Ganymede $= 6 \times 10^{11} \text{ m}$ (closest approach);

  $\alpha'' = 1/2''$, so $(1/2)/(2.06 \times 10^5) = d/(6 \times 10^1 \text{ m})$;

  $d = \dfrac{3 \times 10^{11} \text{ m}}{2.06 \times 10^5} = 1.45 \times 10^6 \text{ m} = 1450 \text{ km}$

9. $W = \text{weight} = GMm/r^{\,2}$, so
  $W_{\text{Jupiter}}/W_{\text{Earth}} = (GM_{\text{Jupiter}}m/r^2{}_{\text{Jupiter}})/(GM_{\text{Earth}}m/r^2{}_{\text{Earth}})$.
  Use Table 8-1 to get numbers.
  $W_{\text{Jupiter}}/W_{\text{Earth}} = 2.5$

10. $v = (3kT/m)^{1/2}$
  The H molecule has two atoms, so
  $m = 2 \times 1.67 \times 10^{-27} \text{ kg} = 3.34 \times 10^{-27} \text{ kg}$.
  Putting in the numbers for k and m, we get
  $v = 111 \times (T)^{1/2} \text{ m/s}$.

  a. $T = 1500 \, K$, $T^{1/2} = 39$, $v = 4.33 \times 10^3 \text{ m/s}$.

  b. $T = 1500 \, K$, so the answer must be the same as in part a.

c. $T = 200\ K$, $T^{1/2} = 14$, $v = 1.55 \times 10^3$ m/s.

d. Escape velocity is discussed in Chapter 4.

$$v\ (esc) = (2GM/r)^{1/2}$$

and the figure is 11 km/s for the Earth. We look at ratios for object (P) and Earth (E):

$$v\ (esc)_{(P)}/v\ (esc)_{(E)} = (M_P/M_E)^{1/2}\ (r_E/r_P)^{1/2}$$

Use the numbers in Table 8-1 and get 11 km/s for Earth, 2.3 km/s for the Moon, and 59 km/s for Jupiter.

The Moon's escape velocity is clearly lower than the hydrogen speed, and the escape speed for Jupiter is clearly much higher. The figure for the Earth is about equal, but top speeds are often three times higher, and the 11 km/s is for the Earth's surface, not the upper atmosphere.

Sample Test Questions

True-False

1. The largest planet is Jupiter. T
2. All gas giant planets in the solar system have several natural satellites. T
3. Jupiter has many light and dark stripes parallel to its rotation. T
4. The dark bands on Jupiter are called belts. T
5. Jupiter's atmosphere is mostly hydrogen and nitrogen. F
6. The upper part of Jupiter's atmosphere is about 133 K. T
7. The lower atmosphere of Jupiter is much colder than the upper atmosphere. F
8. The red spot on Jupiter has existed at most for 100 years. F
9. Jupiter is radiating about two times the energy it receives from the Sun. T
10. Jupiter gets its heat from nuclear fusion. F
11. The synchrotron radiation from Jupiter is at a wavelength of about 1/10 meter. T
12. Jupiter's decameter radio radiation that we receive on Earth depends upon the position of Ganymede with respect to the Earth. F
13. Jupiter's interior is about 60% hydrogen. T
14. The four largest satellites of Jupiter are called Galilean satellites. T
15. Ganymede is larger than Mercury.　T
16. The Galilean satellites of Jupiter all have prograde orbits. T

Multiple Choice

1. The largest planet in the solar system is
   - A. Earth
   - *B. Jupiter
   - C. Saturn
   - D. Uranus
   - E. Pluto

2. Jupiter's atmosphere is mostly hydrogen and
   - *A. helium
   - B. neon
   - C. water
   - D. ammonia
   - E. sulfuric acid

3. The upper atmosphere of Jupiter is about _____ K.
   A. 0   D. 273
   B. 50   E. 670
 *C. 133

4. The lower atmosphere (60 km deep) of Jupiter probably has temperatures near those of Earth but with pressures 10 times Earth's.
 *A. correct
   B. wrong: cooler
   C. wrong: much warmer
   D. wrong: pressures 1000 times Earth's at that depth
   E. wrong: both C and D changes needed

5. Jupiter rotates with a day of about _____ hours.
   A. 1   D. 59
 *B. 10   E. 100
   C. 24

6. Jupiter is warmer than can be expected from the Sun's heat. This extra heat is produced by thermonuclear reactions in the core.
   A. correct
 *B. wrong: The heat is produced by gravitational contraction.
   C. wrong: The heat is produced by radioactive decay.
   D. misleading: It has the temperature expected from solar heating.
   E. misleading: It is cooler than expected from solar heating.

7. Synchrotron radiation is caused by
 *A. electrons moving rapidly in a magnetic field
   B. hydrogen atoms moving rapidly in a magnetic field
   C. thermal heating
   D. Jupiter's moon Io
   E. poorly shielded CBs

8. Jupiter's decimeter radiation is
   A. thermal radiation
   B. from the surface of Jupiter
 *C. synchrotron radiation
   D. both A and C needed
   E. misleading: Jupiter radiates only at decameter wavelengths, not decimeter.

9. Jupiter is emitting radio signals at decameter wavelengths (noise); the strength of these signals is related to the relative position (with respect to the Earth and Jupiter) of one of Jupiter's satellites.
   A. The signals are regular and seem to be intelligent signals.
 *B. Statement is correct as it stands.
   C. Jupiter does not emit radio signals.
   D. The radiation described is at decimeter wavelengths.

10. Jupiter's interior is mostly _____, which can conduct electrons.
   A. level
   B. granite
   C. carbon
   *D. metallic hydrogen
   E. metallic nitrogen

Essay

1. Sketch the disk of Jupiter and any features visible through an Earth-based telescope. Label at least three major items. (See Fig. 11-2.)
2. Discuss the composition of the Jovian moons. (See pp. 231–242.)
3. Discuss the volcanic features of Io. (See pp. 239–242.)
4. Why do the giant planets have massive atmospheres? (See pp. 243–244.)

Word Practice

1. _____ _____ is emitted when very fast electrons move through a magnetic field.

2. _____ rotation is east to west.

3. The _____ _____ of Jupiter is a long-lived storm.

4. The four large satellites of Jupiter are called the_____ satellites.

5. Jupiter's _____ thermal radiation is due to the planet's heat.

6. Jupiter's _____ is about 133 K.

*7. The great Red Spot is often visible on the disk of _____.

*8. Jupiter's satellite _____ is often surrounded by a sodium cloud.

9. Much of Jupiter's interior is _____ _____.

*10. _____ is the planet with the largest number of known satellites (not counting ring particles).

11. The opposite of prograde is _____.

*12. The outer planets are characterized by _____ _____ and large size.

*13. Jupiter's _____ radiation is snychrotron radiation.

14. Jupiter's infrared radiation is _____ radiation.

*15. Jupiter's short-wave radiation is often called _____ radiation.

16. Jupiter is such that it may be hard to say where the surface divides the planet from its _____.

17. Satellites that revolve about their parent body in the same sense as the planets revolve about the Sun are said to revolve _____.

*18. Jupiter's _____ radiation as measured on Earth depends upon the position of Io.

19. _____ satellites revolve counter to the planet's revolution.

*20. Jupiter's long-wave radiation is in the _____ range.

21. The dark lines across the disk of Jupiter are called _____.

*22. The most abundant gas in Jupiter's atmosphere is _____.

23. The bright cloud bands parallel to Jupiter's equator are called _____.

24. Jupiter's _____ is about 60% hydrogen.

25. The rocky matter making up the moons of Jupiter is a black_____ material.

26. The Galilean satellite covered with ice is _____.

Answers to Word Practice

1. synchrotron radiation
2. retrograde
3. Red Spot
4. Galilean
5. infrared
6. temperature
7. Jupiter
8. Io
9. metallic hydrogen
10. Jupiter
11. retrograde
12. low density
13. short-wave
14. thermal
15. decimeter
16. atmosphere
17. prograde
18. long-wave
19. retrograde
20. decameter
21. belts
22. hydrogen
23. zones
24. interior
25. carbonaceous
26. Europa

Chapter 12
The Outermost Planets and Their Moons

Answers to Problems in the Text

1. The atmospheres of Uranus and Neptune are too cold for the organic compounds found in the atmospheres of Jupiter and Saturn. The color of the outer two planets' atmospheres is dominated by a high-altitude methane-rich haze, which absorbs the redder light and reflects the blue light, the same roles played by water vapor in the Earth's atmosphere.
2. The rings of Jupiter, Saturn, and Neptune were probably formed by the fragmentation of moons that moved too close to the planet, or they may be due to material in orbit being unable to collapse into a moon. Both possibilities involve tidal effects. (See p. 252.)
3. The Earth is too warm to keep hydrogen and helium in its atmosphere. Optional Equation VI in Chapter 11 discusses the average velocity, and we previously discussed escape velocity in Optional Equation IV. The gas giants were cooler and hence kept this material; this made them more massive and hence able to keep even more material.
4. The ring system would appear high in the sky, part of it extending far from the sunset side as a bright, narrow (low-latitude) band with a fading on the side opposite sunset. Several moons would probably be visible in the darkening sky.
5. The atmosphere of Titan is mostly nitrogen, with 10% methane and smog. The methane as well as ethane may also exist as both a solid and a liquid on the surface. This provides a good environment for complex organic compounds to form. Thus, Titan's surface is similar to the early terrestrial surface.
6. In both kinds of pass, the craft would be moving rapidly relative to the particles that make up the rings. In the plane, the probe would probably suffer many collisions (remember, 8 km/≈ 1000 miles per hour). However, perpendicular passage would take only 0.1 second, so a collision is unlikely, but there would also be little time for collecting data. Of course, data have now been successfully collected as a probe passed through the ring structure.
7. At mid-latitudes, a yearly cycle would include about four months of day and about two months of 24-hour day as the Earth's axis points, respectively, away from it at right angles to, toward, and again at right angles to the Sun. Seasons would be much more extreme than we now have.
8. Only the orbits of Pluto and Neptune. Of course, the orbits of some asteroids, comets, and Chiron cross planetary orbits.
9. Bode's rule: 4 + one of a series.
   0, 3, 6, 12, 24, 48, 96, 192, 384, 768 divided by 10
   ⇑   ⇑   ⇑   ⇑
   Mercury Mars   Saturn   Pluto
   So, the rule predicts a tenth planet at 77.2 AU. Searches have yielded no large planet there. Snow because it reflects about six times more light than rock does.
10. Since Saturn is much farther from the Sun than the Earth is, we see it only in the full or nearly full phase. Fig. 12-7 had to be taken at a distance about as large or larger than the Sun-Saturn distance.

11. Use Wien's law. Wavelength $= \dfrac{0.0029}{T}$,

    so $T = 2.9 \times 10^{-3}/\text{wavelength} = 2.9 \times 10^{-3}/2 \times 10^{-5}$ m
    $= 1.5 \times 10^2$ K ~ 150 K.

12. $M_{\hbar} = 5.7 \times 10^{26}$ kg

    a. $V_{\text{circ}} = \sqrt{(GM/R)}$; $G = 6.67 \times 10^{-11}$ SI Units

    Take $R = 2.5 \times 10^5$ km $= 2.5 \times 10^8$ m.

    $V_{\text{circ}} = [(6.67 \times 5.7/2.5) \times (10^{-11} \times 10^{26} \times 10^{-8})]^{1/2}$
    $= (1.52 \times 10^8)^{1/2}$ m/s

    $V_{\text{circ}} = 1.23 \times 10^4$ m/s $= 12.3$ km/s

    b. In principle, we just subtract the circular velocity at two nearby radii (1 km).

    $$\Delta V \equiv V_{\text{circ2}} - V_{\text{circ1}} = \sqrt{GM/R} - \sqrt{GM/R}$$
    $$= H(GM)\left(1/\sqrt{R_1} - 1/\sqrt{R_2}\right)$$

This type of calculation is worthwhile only if a calculator is available to carry answers to many significant figures so that the difference is meaningful. Otherwise, the solution is meaningless and we will have to use other mathematical methods. This point is important to make in a mathematically oriented course.

If the students have not had calculus, then to solve this problem we need only the expansion of $(1 + x)^n = 1 + nx$, if x is much less than 1. That is:

Let $R_1 = R_2 + dR$

$1/R = (1/R_2)(1 - dR/R_2)^{-1}$

$(1 + x)^{-1} = 1 - x$ if x is small

$1/R_1 = (1/R_2)(1 - dR/R_2)$

$\sqrt{1/R_1} = \sqrt{1/R_2}\sqrt{1 - dR/R_2}$

$(1 + x)^{1/2} = 1 + x/2$, if x is small,

so $\sqrt{GM/R_1}\ \sqrt{GM/R_2}\ [1 - 1/2(dR/R_2)] = V_2[1 - 1/2(dR/R_2)]$;

hence,

$$|dV| = \frac{1}{2}V_2\frac{dR}{R} = [(1/2) \times 12.3 \text{ km/s}]\ \frac{10^3\text{m}}{2.5 \times 10^8\text{m}}$$
$$= 2.5 \times 10^{-5} \text{ km/s.}$$

Relative velocity is 2.5 cm/s (if catch-up collision occurs).

If the students have had calculus,

$V = (GM)^{1/2}R^{-1/2}$

$$dV = -\frac{1}{2}(GM)^{1/2}R^{-3/2}\ dR = \frac{V}{2R}\ dR$$

13. W = weight = GM m/r^2, so

$$W_{Titan}/W_{Earth} = \left(GM_{Titan}\frac{m}{r^2_{Titan}}\right)/(GM_{Earth}m/r^2_{Earth})$$

$$= \frac{M_{Titan}}{M_{Earth}}\left(\frac{r_{Earth}}{r_{Titan}}\right)^2.$$

Use Table 8-1 to get numbers.
W_{Titan}/W_{Earth} = 0.13

14. Use mean density = mass/volume = $3M/(4\pi^3)$ = 625 kg/m^3. This is less than water (1000 kg/m^3), so any rocky core of Saturn must be small.

Sample Test Questions

True-False

1. All gas giant planets in the solar system have several natural satellites. T
2. There is a red spot in Saturn. F
3. Saturn has a day of between 10^h and 11^h. T
4. The large gap between the outer ring of Saturn and the bright inner ring is called Cassini's division. T
5. Each of Saturn's rings is one solid object encircling the planet. F
6. Saturn's rings are inside the Roche limit for Saturn. T
7. Not counting the rings, Saturn has at least 17 satellites. T
8. Saturn's moon Titan has an atmosphere. T
9. Uranus was discovered by accident. T
10. Uranus has an obliquity of 98°. T
11. Uranus' rotation is prograde. F
12. Saturn is the only planet with rings. F
13. Neptune was discovered by accident. F
14. Pluto is probably an escaped satellite of Neptune. T
15. There might easily be a planet as large as Neptune at 90 AU from the Sun if it lies in the plane of the solar system. F

Multiple Choice

1. The second largest planet in the solar system is
   A. Earth   D. Uranus
   B. Jupiter   E. Pluto
   *C. Saturn

2. Saturn's atmosphere is mostly hydrogen and
   *A. helium   D. ammonia
   B. neon   E. sulfuric acid
   C. water

3. The upper atmosphere of Saturn is about _____ K.
   - A. 0
   - B. 10
   - *C. 100
   - D. 273
   - E. 670

4. The small moons that help confine particles in the F ring of Saturn are called _____ satellites.
   - A. herding
   - B. confining
   - *C. shepherd
   - D. sheep
   - E. ring-moon

5. The broad division or gap between the outer ring of Saturn and the brighter inner ring is called
   - A. Encke's division
   - *B. Cassini's division
   - C. the gap
   - D. the separator
   - E. none of these

6. The particles in Saturn's rings come from
   - A. a condensed gas as Saturn formed
   - B. part of a satellite broken up by a collision
   - C. a comet or asteroid broken apart by tidal forces
   - D. either B or C
   - *E. any of the three, A, B, and/or C

7. One of the largest satellites is Titan, which circles
   - A. Jupiter
   - B. Mars
   - *C. Saturn
   - D. Uranus
   - E. none of these

8. Neither Neptune nor Uranus was known to the ancient astronomers. Uranus was discovered by accident, but Neptune was found by the effects it had on Uranus' orbit.
   - *A. correct
   - B. wrong: The ancient Greeks knew of Uranus but not of Neptune.
   - C. wrong: The second sentence is correct only if the words Uranus and Neptune are interchanged.
   - D. wrong: Both B and C are true.
   - E. wrong: The effects on Pluto's orbit were what led to the discovery of Neptune.

9. We expect the surface of Titan to be covered with liquid
   - A. water
   - B. hydrogen
   - C. helium
   - *D. methane
   - E. sulfur dioxide

10. Of all the outer planets, only Jupiter and Saturn have rings.
    A. correct
    B. wrong: Neptune also has rings.
    C. wrong: Uranus also has rings.
    *D. wrong: Both Neptune and Uranus also have rings.
    E. wrong: Only Saturn and Neptune have rings.

11. Most of the moons of the outer planets are composed only of
    A. light-colored ices
    B. sooty black organic materials
    *C. Both A and B are found alone or together.
    D. No ice is observed in the satellites of these planets.

Problems Involving Optional Mathematical Equations from Other Chapters

1. (Optional Equation I) Cassini's division is 1500 km wide. When Saturn is closest
   to Earth, it is about 9 AU away from us. Which telescope listed below would show
   Cassini's division?
   *A. a telescope with a resolution of 0."1
   B. a telescope with a resolution of 1."0
   C. a telescope with a resolution of 10"
   D. both A and B
   E. all three, A, B, and C

2. (Optional Equation V) Uranus has a mass about 15 times the Earth's and a radius
   about four times the Earth's. How many times stronger than the Earth's gravitational
   field is that of Uranus at its surface (approximately)? That is, surface gravity on
   Uranus = (approximately) _____ surface gravity on the Earth.
   A. 1/10   D. 9
   *B. 1   E. 56
   C. 4

3. (Optional Equation IV) The surface temperature of Pluto is about 50 K. At what
   wavelength is the most thermal radiation emitted (not reflected)?
   A. 3 cm   D. 5.8×10^5 m
   B. 5.8 m   *E. 5.8×10^{-5} m
   C. .145 m

Essay

1. Sketch the disk of Saturn and the surrounding region. Draw any features visible
   through an Earth-based telescope. Label at least three major items, including
   Cassini's division. (See Fig. 12-2.)
2. List the three possible sources of the particles of the rings of Saturn. (See p. 252.)
3. Describe how the rings of Uranus were discovered. (See p. 263.)
4. Discuss the typical composition of the satellites of the outer planets. (See p. 270.)
5. Discuss the atmospheres of the gas giants. (See pp. 269–270.)

1. _____ has rings easily visible through a telescope.

2. The planet _____ was discovered by its gravitational effects on the orbit of Uranus.

3. _____ division is a prominent gap in the rings of Saturn.

4. The outermost planet of the solar system is_____.

5. _____ _____ system is very thin and composed of many small particles.

6. _____ atmosphere is similar to Jupiter's.

7. _____ _____ may be related to the peculiar orbit of Triton.

8. The first planet discovered in recorded history is _____.

9. _____ _____ can help confine rings to a narrow zone.

10. _____ _____ _____ begins at the edge of the rings with five small satellites.

11. The satellite _____ of Saturn has clearly broken apart in the past and has reassembled.

12. Saturn's satellite _____ seems to have a structure intermediate between Europa and Ganymede of Jupiter.

13. _____, _____, and _____ are three small moons of Saturn just inside the orbit of Titan.

14. The moon _____ of Saturn has an atmosphere.

15. _____ is an irregular moon of Saturn in chaotic rotation.

16. _____ _____ occurs when the rotation rate and direction of a body axis changes irregularly.

17. _____ is a satellite of Saturn with one light and one dark side.

18. Saturn's satellite _____ is a small black-surfaced object in retrograde orbit.

19. The _____ of a planet is the deviation of the axis of rotation from being perpendicular to the orbital plane.

20. Uranus' _____ were discovered only in 1977 when they passed in front of a star.

21. _____ satellites were mostly discovered by Voyager.

22. _____ is the innermost of the five largest satellites of Uranus. It has a highly fractured surface.

23. _____ is the largest satellite of Neptune. It may have a thin atmosphere.

24. _____ is the only known satellite of Pluto.

25. _____ is the term used to refer to a possible planet in orbit beyond Pluto.

Answers to Word Practice

1. Saturn
2. Neptune
3. Cassini's
4. Pluto
5. Saturn's ring
6. Saturn's
7. Pluto's origin
8. Uranus
9. shepherd satellites
10. Saturn's satellite system
11. Mimas
12. Enceladus
13. Tethys, Dione, and Rhea
14. Titan
15. Hyperion
16. chaotic rotation
17. Iapetus
18. Phoebe
19. obliquity
20. rings
21. Uranus'
22. Miranda
23. Triton
24. Charon
25. "Planet X"

Chapter 13
Comets, Meteors, Asteroids, and Meteorites

Answers to Problems in the Text

1. Still being too cold to vaporize, it would not release a large gas cloud. Hence, it would be too small and faint to be seen.
2. Yes. It applies to all bodies orbiting around the Sun or in fact all small bodies orbiting around a much more massive body. (Hence, the term "Keplerian motion.")
   If $a = 10^5$ AU; $p^2 = a^3 = 10^{15}$;

   $$p = \sqrt{10} \times 10^7 = 3.2 \times 10^7 \text{ years} = 32 \text{ million years.}$$

3.

Item	Useful Form	Nonuseful Form
Time	clockhour (e.g., 8:37 PM EST)	vague—at night (e.g., sometime last night)
Size	angular or relative to landmarks (e.g., 2° or as large as the width of the mountain visible from my house)	imprecise measurement (e.g., as large as a horse, as big as a 747, biggest thing I ever saw)
Brightness	relative to stars (e.g., between Sirius and Venus in brightness)	absolute, untestable, (e.g., brightest thing I ever saw)
Motion	relative to stars (e.g., passed between Mars and Jupiter in three minutes)	absolute (e.g., moved 100 mph or as fast as a plane)
Location	relative to sky objects (e.g., passed through Orion)	vague (e.g., overhead)

Make sure of location and time of fall. Notify nearest science museum, geology department, or astronomy department. Return and remain near fall. Protect it from being disturbed. Notify an astronomer through his or her textbook publisher if no local authority seems concerned or competent.

4. Meteors: glow as meteoroid hits atmosphere; fine debris of comets.
   Comets: gaseous pieces of asteroids kicked out into cometary orbits.
   Zodiacal light: microscopic meteoroid dust left in solar system (remnants of comets and asteroids? primordial?).
   Asteroids: small bodies mostly in orbit between Mars and Jupiter; some elsewhere; debris of solar system formation. Some objects listed as asteroids may be burned-out comets.
   Meteorites: probably pieces of asteroids (possibly comets) that survive as meteors and hit the ground.

5. An icy body some miles across. Icy surface or perhaps covered with a dusty layer of soil dislodged as ices sublime. It might not really have a surface as we think of it. Entire sky and region about the astronauts would glow from the vaporized material if in the inner solar system. Gravity would be negligible. Short-period comets would have speeds closer to the Earth's velocity and would more likely lie in the plane of the solar system and hence be easier to match orbits with.

6. Neglect the fact that the Earth would accelerate particles as they got very close to it. All numbers are thus upper limits: speed = 15 km/s.

   a. time = distance/velocity; distance = 15×10^6 km; time = 10^6 s = 11.6 days.

   b. This is not much time to move many people. If the object were going to hit a body of water (which they do in five out of six cases), people could be evacuated to higher ground. The process of political decision might take several days, so even less time is left.

   c. distance = 3.81×10^5 cm; time = distance/velocity = 2.56×10^4 s = 7.1 hours.

   d. distance = 100 km; time = distance/velocity = 6.7 s.

   Note: Many students may be helped by dimensional analysis, so you could set the problem up as

   $$\text{time} = m \ \times \ \frac{s}{m} = m/(m/s);$$

   that is, time = distance/velocity.

7. Very small. Yes: Survivors might have reported a finger of God striking the area or destruction raining down from heaven because of the (local) gods. If the object fell into the sea, the surrounding coasts would have been flooded by tsunamis. This could have resulted in flood myths.

Advanced Problems

8. Since the comet is on a parabolic orbit, we may assume it is traveling at the speed needed to escape from the Sun when it is at closest approach. The Earth is in nearly circular orbit, so the Earth's speed is that of circular orbit. Thus, $V_{comet} =$ $\sqrt{2}\, V_{circular}$ as we saw in Chapter 4 (Optional Equation III). So the comet's speed is either $(\sqrt{2} - 1)$ or $(\sqrt{2} + 1)$ times the Earth's speed. Earth velocity in orbit is

   $$V \ = \ \frac{2\pi R}{period} = \frac{2\pi \times 1.5 \times 10^{11} m}{3 \times 10^7 s} \approx 3 \times 10^4 m/s$$

   Therefore, 1.2×10^4 m/s or 7.2×10^4 m/s.

9. $\alpha'' = 1''$, $d = 2$ km $= 2 \times 10^3$ m

   $\alpha''/(2.06 \times 10^5) = d/D \Rightarrow D = d \times 2.06 \times 10^5/\alpha''$

   $D = 2 \times 10^3 \times 2.06 \times 10^5$ m $= 4.06 \times 10^5$ km

   This distance is slightly more than the Moon's distance from the Earth; however, it would have to be closer to the Moon's orbit for us to be able to distinguish its shape.

10. Mass $= 4/3\pi r^3 p$ $= (4/3) \times 3.14 \times (10^2$ m$)^3 \times 8000$ kg/m^3

    $= 3.35 \times 10^{10}$ kg

    Value $=$ mass $\times$ \$.90/kg $= \$3.02 \times 10^{10} = \$30$ billion

Thus, investors (or the government) could spend up to about \$30 billion to reach and mine this body if the price remained at this level. Future shortages of many essential substances may make the possible future return even greater.

11. Volume $= \frac{4}{3}\pi r^3 = \frac{1}{6}\pi(\text{diameter})^3 \Rightarrow \dfrac{\text{volume of Earth}}{\text{volume of asteroid}}$

$$= \left(\frac{\text{dia} \oplus}{\text{diameter}}\right)^3 = \left(\frac{1.2 \times 10^7}{1.2 \times 10^5}\right)^3 = 10^6$$

It would take about 1 million.

Sample Test Questions

True-False

1. The cosmic debris floating between the planets is known as the meteoritic complex. T
2. Comets are rocky-metallic worlds ranging from a few hundred meters or less up to 1000 km in diameter. F
3. Interplanetary rocks that collide with the Earth's atmosphere and survive to fall to the ground are termed meteors. F
4. The original small, preplanetary bodies of the solar system are known as planetesimals. T
5. Comets are omens of things to come. F
6. Tycho Brahe was the first astronomer to determine that comets were more distant than the Moon. T
7. The Oort cloud is a swarm of millions of comets at 50 AU from the Sun. F
8. The Oort cloud is a swarm of millions of comets at 50,000 AU from the Sun. T
9. Nearby stars cause comets to leave the Oort cloud. T
10. Most meteors must be small bits of debris scattered from comets. T
11. The asteroid belt is the region between Mars and Saturn. F
12. Asteroids that cross the Earth's orbit are called Apollo asteroids. T
13. The smaller the meteorite, the rarer it is. F
14. There are four basic types of meteorites. F
15. The meteorite parent bodies formed 4.6 billion years ago. T
16. The zodiacal light is a faint glowing band of light diametrically opposite the Sun. T

Multiple Choice

1. Icy worlds a few kilometers across are
   A. parent bodies
   B. meteoroids
   C. meteors
   *D. comets
   E. asteroids

2. A small nebulosity with one or more bright spots and an attached flowing streak extending several degrees across the sky is
   A. a meteor
   B. an asteroid
   C. a planet
   *D. a comet
   E. a galaxy

Consider the sketch of a comet:

3. The comet nucleus is denoted by __A__.

4. The comet tail is denoted by __C__.

5. The comet head is denoted by __D__.

D. A and B together
E. B and C together

6. The dust tails of comets point
   A. toward the Sun
   *B. away from the Sun
   C. toward the Earth
   D. toward Jupiter
   E. toward their direction of travel

7. Comets may have several tails, but they are of only two types.
   *A. correct
   B. wrong: only two tails at most
   C. wrong: only one tail
   D. wrong: at least five types
   E. wrong: both B and D

8. Most comets are believed to come from other nearby stars.
   A. correct
   B. wrong: ejected from the Sun
   *C. wrong: true members of the solar system
   D. wrong: from other galaxies
   E. wrong: astronomers have no idea

9. The comet model that suggests chunks of frozen gases several kilometers across, laced with bits of silicate dust, is called the _____ model.
   A. Oort
   B. frozen sherbet
   *C. dirty iceberg
   D. frozen lace
   E. silicate

10. Tails of comets are
    A. caused by ionization of the Earth's atmosphere as the comet enters it
    B. dust and gas captured by the comet
    C. the jet exhausts of the engines that make the comets move
    *D. dust and gas blown off the comet by the solar wind
    E. misleading: Comets do not have tails.

11. Most meteors are
    *A. small bits of debris scattered from comets
    B. small comets that never had long tails
    C. pieces of Jupiter
    D. pieces of Saturn's rings
    E. none of these

12. The asteroids that lie in two swarms 60° ahead of and behind Jupiter are the asteroids
    A. Apollo
    B. Jupiter
    C. Lagrange
    *D. Trojan
    E. Greek

13. The asteroids are small bodies, most of whose orbits lie near the plane of the ecliptic between the orbits of Jupiter and Mars.
    *A. correct
    B. wrong: The asteroids are usually larger than the Moon.
    C. wrong: The orbits of the asteroids lie between Jupiter and Saturn.
    D. wrong: There is no preferred plane for asteroid orbits.
    E. none of the above

14. The points 60° ahead of and behind a planet in its orbit are called the _____ points. At these points, the planet and the Sun permit other particles to exist in a stable orbit.
    A. Apollo
    *B. Lagrangian
    C. Trojan
    D. stable
    E. Oort

15. The _____ are a stony type of meteorite that has never been subjected to very much heat.
    *A. carbonaceous chondrites
    B. achondrites
    C. chondrites
    D. irons
    E. stony-irons

16. The chondrites come from the _____ of the parent body of meteorites.
    *A. outside
    B. core
    C. melted metals
    D. radioactive regions
    E. zodiac

17. Most meteorites became solid objects 1.2 billion years ago as parts of parent bodies. These parent bodies broke up as late as a few million years ago.
    A. correct
    B. wrong: Parents broke up 1.2 billion years ago.
    *C. wrong: Most became solid objects 4.6 billion years ago.
    D. wrong: Both B and C changes are needed.
    E. misleading: There is no indication that meteorites ever were parts of larger bodies.

*18. (Optional Equation I) Ceres is about 1000 km in diameter and can be as close as 1.77 AU from the Earth. Which of the telescopes below could resolve Ceres?
  A. a telescope with a resolution of 0."07
  B. a telescope with a resolution of 0."7
  C. a telescope with a resolution of 7"
*D. both A and B
  E. all three, A, B, and C

*19. (Optional Equation II) Ceres has an estimated mass of 1.2×10^{21} kg and a radius of 500 km. Thus, it has about 2×10^{-4} times the Earth's mass and about 0.079 times its radius. So the gravitational force of Ceres equals (approximately) _____ times the gravitational force on Earth.
  A. 3×10^{-7} D. 3.2
*B. 3.2×10^{-2} E. 965
  C. 0.32

*20. (Optional Equation III) Vesta has an estimated mass of 2.4×10^{20} kg and a radius of 269 km. What is the approximate escape velocity from Vesta?
  A. 0.035 km/s   D. 0.24 km/s
*B. 0.35 km/s   E. 0.024 km/s
  C. 3.5 km/s

Essay

1. Distinguish among meteoroids, meteors, and meteorites, commenting on their origins. (See p. 273.)
2. Sketch a comet, labeling nucleus, head, and tail. (See Multiple Choice questions 3–5.)
3. Discuss the cause of zodiacal light. (See p. 293.)
4. What can be learned from a study of meteorites? (See pp. 291–292.)
5. Why are the small bodies in the solar system so useful in teaching us about the formation of the solar system? (See pp. 295–296.)

Word Practice

1. Occasionally large meteors occur. Such a large meteor is called a(n)

   _____.

2. Whipple has proposed the _____ _____ _____ of comet nuclei—chunks of frozen gases laced with bits of silicate dust.

3. _____ are icy worlds of few kilometers across when they are far from the Sun.

4. _____ are small bits of debris in space.

5. _____, _____, and _____ of all sizes are all debris left over from the origin of the solar system 4.6 billion years ago.

6. A(n) _____ meteorite is made of mixed fragments of different types.

7. The believed cloud of comets at 50,000 AU from the Sun is called the _____ _____.

8. The best samples of _____ _____ _____ are the carbonaceous chondrites.

9. _____ are the millimeter-scale spherical silicate inclusions in chondrites.

10. The particles blown off the Sun are referred to as the _____ _____.

11. _____ are meteorites most nearly like terrestrial igneous rocks.

12. _____ _____ are meteorites composed of the most primitive, least altered material.

13. A meteor _____ occurs when many meteors are seen in a short time interval, all coming from approximately the same place in the sky.

14. Meteorites with a large metal content are called _____ meteorites.

15. The _____ asteroids cross the Earth's orbit.

16. The _____ _____ is sunlight reflected off the microscopic dust grains in the solar system.

17. A meteor is seen that seems to strike the ground. Later, _____ are found there.

18. Asteroids that cross the orbit of Mars but not the Earth are called _____ _____ _____.

19. The first comet predicted to return was _____ Comet.

*20. _____ was the first asteroid discovered.

21. The _____ asteroids lie in two swarms 60° ahead of and behind Jupiter.

22. _____ are what is left of meteoroids that reach the ground.

23. _____ appear as a glowing streak across part of the sky. They remain for the night, slowly moving with respect to the stars.

24. The nucleus and the coma of a comet make up the _____.

25. _____ _____ are composed of microscopic particles blown off comets and now freely orbiting the Sun.

26. _____ meteorites are mixtures of silicates and metals.

27. The _____ points of Jupiter are the points 60° ahead of and behind Jupiter in its orbit.

28. _____ _____ comets take only a few decades for a complete orbit.

29. Most of the asteroids lie in the _____ _____ that is between Mars and Jupiter.

30. The _____ of a comet is the starlike point in the comet head.

31. _____ were small preplanetary bodies in the early solar system.

32. A glowing streak is seen across the sky and disappears in a few seconds. _____ produce such an effect.

33. Meteorites broke off of a larger body called the _____ _____.

34. The _____ belt lies between Mars and Jupiter.

35. The comet _____ is a faint glow extending from the coma.

36. There are _____, iron, and _____-iron meteorites.

Answers to Word Practice

1. fireball
2. dirty iceberg model
3. comets
4. meteoroids
5. comets, meteorites, asteroids
6. brecciated
7. Oort cloud
8. primitive planetary material
9. chondrules
10. solar wind
11. achondrites
12. carbonaceous chondrites
13. shower
14. iron
15. Apollo
16. zodiacal light
17. meteorites
18. Mars-crossing asteroids
19. Halley's
20. Ceres
21. Trojan
22. meteorites
23. comets
24. head
25. dust trails
26. stony-iron
27. Lagrangian
28. short-period
29. asteroid belt
30. nucleus
31. planetesimals
32. meteors
33. parent body
34. asteroid
35. tail
36. stony, stony

Chapter 14
Origin of the Solar System

Answers to Problems in the Text

1. This is expected, from the basic laws of physics, as a natural evolutionary behavior. Thus, for planets of terrestrial size and temperatures, there is no need to explain the lack of hydrogen as an initial condition or primeval process.

2. Rocks on the Earth have been subjected to erosion by water and cooking by plate tectonics. The meteorite rocks, with rare exceptions, are closer than Earth rocks to the form they had when the material condensed. Meteorites have somewhat different chemistries because they were formed in a different part of the solar nebula.

3. The planets in the hotter inner solar system could not hold hydrogen. That is, we would expect the inner planets to have a smaller percentage of H, N, and C than the outer planets and a correspondingly larger percentage of Al, Ti, Ca, Fe, Ni, and Si. See Table 14-2. Even the compounds formed can be explained. For example, Venus and Mercury were too hot for much water ever to form. Compared to the Earth, Mercury is richer in iron and Mars is poorer. The other planets have abundant H-rich ices.

4. The planets Mercury, Venus, Earth, and Mars contain much less "ices" than Jupiter, Saturn, Uranus, and Neptune, so we expect that the "inner boundary" for condensation of ices was between Mars and Jupiter. A visit to the asteroid belt may tell us where between these two planets the boundary was.

5. The whole idea of a condensing gas and dust cloud might have to be thrown out if the orbits were noncircular and inclined. Studies show that a cloud of dust and gas should flatten to a disk shape, giving rise to planets with circular orbits in a plane.

6. See entries in Table 14-1. The disk shape of the solar system. Meteorites bear evidence of accumulation and collisions of rock material. Is the distribution of minerals in the right order from the Sun? Do asteroids, meteorites, and craters all imply the existence of planetesimals?

7. A solar system can start with a collapsing dust and gas cloud with elements already existing in the percentages observed in the solar system. Theories of the origin of the universe must explain where all matter comes from and why.

8. Most meteorites and Earth and Moon samples we have examined were formed from gases at that date. Radiometric dating shows that the formation interval from dust to planet was probably well under 100 million years. Much has happened to the material of the Earth since that date, but we still know that it became solid then.

9. You could say that the world began as a cloud of dust. The same thing that can make you fall down and hold you down (which we call gravity) brought the cloud together, making the galaxies, their stars, and all the planets.

10. At 100 K, the maximum radiation occurs at a wavelength W, given by $W = 0.00290/T = 2.9 \times 10^{-6}$ m, a much larger wavelength than red light—infrared (2900 nm). Use an infrared telescope to look at the 2.9×10^{-6} m range for bright (at this wavelength) objects of very small angular size.

11. 1 AU $= 1.5 \times 10^{11}$ m

    1 year $= 365 \times 24 \times 60 \times 60 = 3.16 \times 10^7$ s

    Note: It is handy to remind students that 1 year $\approx \pi \times 10^7$ s.

    Circumference $= 2\pi r$

    $$\text{velocity} = \frac{2\pi r}{\text{period}}, \text{ or use } V = \sqrt{GM_\odot/R}$$

    For Earth

    $$\text{velocity} = \frac{2 \times 3.14 \times 1.5 \times 10^1}{3.16 \times 10^7} = 2.98 \times 10^4 \text{ m/s}$$

    Earth orbital

    velocity $= 29.8$ km/s,

    so 0.1% is 0.298 km/s or about 300 m/s. For Pluto, use $p^2 = r^3$ to get p with $r = 29.4$ AU. Then

    $$r^3 = 5.5 \times 10^4 = p^2, p = 2.35 \times 10^2 \text{ years}$$

    $$\text{period} = 235 \times 3.16 \times 10^7 = 7.42 \times 10^9 \text{ s}$$

    $$\text{velocity} = \frac{6.28 \times 39.4 \times 1.5 \times 10^{11}}{7.42 \times 10^9} = 5 \times 10^3 \text{ m/s}$$

    Pluto orbital velocity = 5 km/s, so 0.190 is 0.05 km/s = 50 m/s.

    Alternative Method I: $v = 2\pi r/p$; $p^2 = r^3$.

    Then V in cm/s is $V =$

    $$\frac{2\pi (r)_{AU} \times 1.5 \times 10^{11}}{(P)_{yr} \times 3.16 \times 10^7}$$

    $$= 2.98 \times 10^4 \frac{r \text{ AU}}{P \text{ yr}} \text{ m/s}$$

    $$= \frac{2.98 \times 10^4}{\sqrt{r \text{ AU}}} \text{ m/s}.$$

    Alternative Method II: Use $V_{circ} = \sqrt{GM/R}$, where M = mass of Sun.

12. To keep gases, the average velocity must be less than one-fourth the escape velocity.

    a. Let density P be 1000 kg/m^3, $T = 500$ K, and diameter $D = 38,000$ km. Then

    mass $= M = (4/3) \pi (D/2)^3 \times P = \pi P D^3/6$.

    Using the given numbers, we get $M = 2.87 \times 10^{25}$ kg.

    From Chapter 4, we have

    $$v_{escape} = (2GM/R)^{1/2}$$

    $$v = (2 \times 6.67 \times 10^{-11} \times 2.87 \times 10^{25}/1.9 \times 10^4)^{1/2}$$

    $$= 14.2 \text{ km/s}.$$

    From Chapter 11,

    $$v_{typical} = (3kT/m)^{1/2}.$$

103

Since we have hydrogen molecules,

$$v_{typical} = (3 \times 1.38 \times 10^{-23} \times 500/3.34 \times 10^{-27})^{1/2}$$
$$= 2.5 \text{ km/s}$$
$$4 \times v_{typical} = 10 \text{ km/s} < 14.2 \text{ km/s}$$

b. Use Table 8-1—$M_{Earth} = 5.98 \times 10^{24}$ kg. The mass used is less than 8–30 times the Earth's mass but clearly bigger than the Earth's mass.

c. We see that the gas giants are gas giants because they have large rocky cores. The Earth, on the other hand, is a rocky body because it is too small to retain the original hydrogen and possibly because it formed at a higher temperature.

Sample Test Questions

True-False

1. Evolutionary theories tend to explain most things as due to relatively gradual processes. T
2. Astronomers now believe that catastrophic theories best explain the origin of the solar system. F
3. A theory of the origin of the solar system must explain why the planets differ in their composition. T
4. A theory of the origin of the solar system must explain Kepler's laws. F
5. The first evolutionary theory was proposed by René Descartes in 1644. T
6. The principle of conservation of angular momentum predicts that a spinning body will rotate faster if its mass moves closer to the axis of rotation. T
7. The nebula that surrounded the contracting Sun is called the solar nebula. T
8. One of the first elements to condense out of the solar nebula was silicon (Si). F
9. The smallest particles, such as dust grains, left after planetary formation were possibly carried away from the solar system by solar radiation and gas. T
10. The most abundant element in the solar system is hydrogen. T
11. It would be easy to develop a solar system cosmology if the solar system consisted of planets orbiting the Sun in randomly inclined circular orbits, both prograde and retrograde. F

Multiple Choice

1. Theories that explain things through sudden or uncommon events are
_____theories.

    A. nebula   D. condensation
*B. catastrophic   E. sudden
    C. evolutionary

2. Any theory of the origin of the solar system must explain
   A. why planetary orbits are nearly circular
   B. why the planet-satellite systems resemble the solar system
   C. why Kepler's laws hold
  *D. both A and B
   E. all three, A, B, and C

3. Any theory of the origin of the solar system must explain
   A. why planetary orbits are nearly circular
   B. why the planets differ in composition
   C. why meteorites differ in chemical and geological properties from all known
       planetary and lunar rocks
   D. both A and B
  *E. all three, A, B, and C

4. The _____ explains both why the planets lie nearly in the same plane and
   why this is the plane of the Sun's rotation.
  *A. conservation of angular momentum
   B. theory of Helmholtz contraction
   C. condensation sequence
   D. conservation of parity
   E. retrograde rotation of Venus

5. An example of conservation of angular momentum is
   A. figure skaters slowing down their spin when they pull in their arms
  *B. figure skaters speeding up their spin when they pull in their arms
   C. a car going through a 90° turn at high speed
   D. what happens when you stand on a skateboard and throw a ball
   E. none of these

6. The collapse of the solar nebula when its atoms were far apart is technically called
   A. Roman falling
   B. catastrophic contraction
  *C. free-fall contraction
   D. Helmholtz contraction
   E. none of these

7. The contraction of the solar nebula in which the shrinkage is slowed by outward
   pressure is
   A. pressure collapse
   B. catastrophic contraction
   C. free-fall contraction
  *D. Helmholtz contraction
   E. none of these

8. A cloud in space that is later incorporated into a planet is called
   A. a cirrus cloud
   B. a grain
   C. a planetesimal
   D. a protoplanet
   *E. none of these

9. The order in which gases condensed out of the solar nebula is termed the
   A. solidification order
   *B. condensation sequence
   C. condensation order
   D. Helmholtz order
   E. Alfvén sequence

10. One of the first elements to condense from the solar nebula was
    A. silicon (Si)
    B. hydrogen (H)
    C. iron (Fe)
    *D. calcium (Ca)
    E. none of these

11. Of the nine elements considered in the condensation sequence in the solar nebula, one
    of the last to condense was
    A. aluminum (Al)
    B. iron (Fe)
    *C. carbon (C)
    D. silicon (Si)
    E. none of these

12. Intermediate bodies from which the planets were assembled are called
    A. grains
    B. nebulae
    *C. planetesimals
    D. intermediate planets
    E. new planets

13. The most abundant element in the solar system is
    A. potassium (K)
    B. sulfur (S)
    C. helium (He)
    D. oxygen (O)
    *E. none of these

14. The second most abundant element in the solar system is
    A. aluminum (Al)
    B. iron (Fe)
    *C. helium (He)
    D. silicon (Si)
    E. none of these

15. The probable method of slowing the Sun's rotation and transforming the angular
    momentum to the planetary material was
    *A. magnetic braking
    B. angular transfer
    C. viscous flow
    D. catastrophic collision
    E. none of these

*16. (Optional Equation III) If a body is in circular motion about a point the distance of which is far greater than the body's size, the body's angular momentum is mvr, where m is the mass, v its velocity, and r the radius of its orbit. Jupiter has a mass of 1.9×10^{27} kg and an orbital radius of 5.2 AU (1 AU = 1.5×10^{1} m). What is the approximate angular momentum of Jupiter in SI units?

  *A. 6×10^{41} D. 2×10^{98}

   B. 6×10^{37} E. 13.5

   C. 4×10^{17}

*17. (Optional Equation IV) Silicate materials condense out of the solar nebula at a temperature of about 1300 K. At that temperature, what would be the approximate wavelength at which the most thermal radiation would be emitted from the resulting mineral grains?

   A. 6.67×10^{-8} m   *D. 2.23×10^{-6} m

   B. 3.77×10^{-2} m   E. 2.06×10^{5} m

   C. 2.23×10^{-5} m

Essay

1. Distinguish between catastrophic theories and evolutionary theories. (See p. 300.)
2. List six observational items that must be explained by a theory of the origin of the solar system. (See Table 14-1.)
3. Give an example of conservation of angular momentum from your everyday experience. (See p. 301.)
4. List three items of evidence for the past existence of planetesimals. (See pp. 304–305.)
5. Discuss the chemical compositions of the planets. (See pp. 309–310.)

Word Practice

1. A(n) _____ is a cloud in space.

2. Microscopic solid particles in space are often called _____.

3. A(n) _____ theory explains things through sudden events.

4. _____ of angular momentum explains why figure skaters spin faster when they pull in their arms.

5. _____ _____ was proposed by H. Alfvén to account for angular momentum transfer within the solar nebula.

6. The intermediate bodies from which the planets were assembled are called

   _____.

7. The _____ _____ is the order in which particles condensed out of the solar nebula.

8. _____ are the bodies intermediate in size between grains and protoplanets.

9. The solar nebula began to collapse in _____ _____.

*10. Catastrophic theories require _____ events.

11. The _____ was the first body to form from the solar nebula.

*12. The third entry in the condensation sequence is _____.

*13. Atoms interacting in a gas cause _____.

*14. Meteorite _____ strongly supports the condensation theory.

15. Magnetic braking explains the transfer of _____ _____ within the solar nebula.

16. When the solar nebula contraction is slowed by outward pressure, we call the contraction _____ _____.

*17. Minerals containing the element _____ formed the first stony material to condense from the solar nebula.

18. The _____ _____ surrounded the forming Sun.

19. The _____ grew into our Sun by absorbing most of the solar nebula.

Answers to Word Practice

1. nebula
2. grains
3. catastrophic
4. conservation
5. magnetic braking
6. planetesimals
7. condensation sequence
8. planetesimals
9. free-fall contraction
10. sudden
11. protosun
12. silicon
13. pressure
14. composition
15. angular momentum
16. Helmholtz contraction
17. silicon
18. solar nebula
19. protosun

Chapter 15
The Sun: The Nature of the Nearest Star

Answers to Problems in the Text

1. Accepting the age of the Sun as 4.6 billion years, we know of no process that could have so long provided the energy for the Sun except nuclear fusion. From the known measurements of nuclear properties and the behavior of gases, plus our knowledge of the Sun's surface temperature, we can calculate approximate internal temperatures and pressure. These predict that nuclear reactions must be occurring in the Sun's hydrogen.

2. One-fourth the apparent solar rotation period: $27\overset{d}{.}75/4 = 6\overset{d}{.}83$. One apparent solar rotation period: $27\overset{d}{.}3$. It is unnecessary to correct for the Earth's motion if one uses the apparent rotation.

3. The number rises and falls every 11 years, but because the magnetic polarity is reversed every cycle, only after two cycles of rising and falling do all properties (polarity and numbers) repeat themselves.

4. Assume that the time required to make the observation is shorter than the interval of observation and that we are talking about the whole Sun, not just one spot.
   a. Changes in granulation spicules and supergranulation; prominences would have moved.
   b. Flares and solar rotation.
   c. Number of sunspots and location with respect to equator.
   d. Solar magnetic field and polarity of sunspots.
   e. Perhaps none, although some astronomers have speculated that known climatic changes on Earth with a 10-year time scale may reflect solar variations of that period.
   f. Nine billion years in the past the Sun would not yet have formed; nine billion years in the future the Sun will have consumed most of its available hydrogen. As later chapters show, its structure must change. We expect it to be a white dwarf by then. We would have missed the red giant stage, which occupies a much shorter period.

5. A radio disturbance is caused by electromagnetic radiation, which takes only $8\overset{m}{.}5$ to reach the Earth. An aurora is caused by particles trapped in the Earth's magnetic field. Particles released by solar flare take days to reach the Earth.

6. The Sun is 1.99×10^{30} kg; Jupiter 1.9×10^{27} kg; Saturn 5.69×10^{26} kg; the rest of the planets combined are negligible, so the planets $= 2.5 \times 10^{27}$ kg and

$$\frac{\text{mass of Sun}}{\text{mass of planets}} = \frac{1.99 \times 10^{30}}{2.5 \times 10^{27}} = 800.$$

We might liken the solar system to a bar of Ivory Soap: 99.9% Sun.

7. Hydrogen, helium.

8. See Chapter 14. The Earth is too small and warm to keep its free hydrogen and helium.

9. It will use up its available hydrogen.

10. Only at the center are the pressures and temperatures high enough for fusion to occur.

11. a. $\dfrac{\alpha''}{2.06 \times 10^5} = d/D$; $d = 1.28 \times 10^7$ m

$\qquad\qquad\qquad\quad = $ diameter of Earth

$\quad D = 1.5 \times 10^{11}$ m $- 1.4 \times 10^9$ m $= 1.5 \times 10^{11}$ m

$\qquad = $ distance to Sun's surface

$\quad \alpha'' = \dfrac{1.28 \times 2.06}{1.5 \times 10^{11}} \times 10^5 \times 10^7 = 1.76 \times 10 = 17''.6.$

  b. The answer is obviously the same as for part a.

12. $V_{esc} = \sqrt{2}\,V_{circ}$

$\quad V_{circ} = \sqrt{GM/R}$; $G = 6.67 \times 10^{-11}$; $M_\odot = 1.99 \times 10^{30}$ kg

$\quad R_\odot = 6.96 \times 10^8$ m

$\quad V_{esc} = [(2 \times 6.67 \times 1.99/6.96) \times 10^{-11} \times 10^{30}/10^8]^{1/2}$

$\qquad = (3.82 \times 10^{11})^{1/2}$ m/s

$\quad V_{esc} = 6.17 \times 10^5$ m/s $= 617$ km/s

13. Wien's law: $W = 0.0029/T$, so $T = 0.0029/W$, where $W = 5.1 \times 10^{-7}$ m, so

$\quad T = \dfrac{29 \times 10^{-4}}{5.1 \times 10^{-7}}$ K $= 5700$ K.

14. a. $v = (3kT/m)^{1/2}$ with $T = 2 \times 10^6$ K. We use the mass for a hydrogen atom or ion as the appropriate value for this problem.

$\quad v = (3 \times 1.38 \times 10^{-23} \times 2 \times 10^6/1.67 \times 10^{-27})^{1/2} = 2.2 \times 10^6$ m/s

  b. From problem 12, $v_{escape} = 617$ km/s for the Sun. Thus, the typical velocity is less than the escape velocity; however, three times the typical velocity is greater than the escape velocity. We expect hydrogen ions and atoms to escape from the Sun.

Sample Test Questions

True-False

1. The Sun appears to rotate only because the Earth is revolving about the Sun. F
2. The study of the colors of light emitted by objects is called spectroscopy. T
3. The array of colors in a light beam, ordered according to wavelength, is called white light. F
4. A glow consisting of a mixture of all colors is called an emission line. F
5. A gas at very high pressure will produce bright emission lines. F
6. A hot gas may produce both a continuous spectrum and dark absorption lines at the same time. T
7. The bright red line emitted by hydrogen under the proper conditions is called the hydrogen alpha line. T
8. The element most abundant in the Sun is oxygen. F
9. Coal burning is an example of a chemical reaction. T
10. The Sun produces its energy by a nuclear reaction called the proton-proton chain. T
11. The Sun's central region is called the corona. F

12. Throughout most of the Sun's volume, energy moves primarily by conduction. F
13. The ability of a gas to obscure light passing through it is called the opacity of the gas. T
14. A device that artificially eclipses the Sun and allows study of the corona and the chromosphere is named an eclipograph. F
15. Sunspots are associated with magnetic fields. T
16. A gas with many ions is called a plasma. T
17. The solar cycle is 11 years long. F
18. Sunspots often appear in pairs. T
19. An outrush of gas from the Sun that reaches far beyond the Earth is called the solar wind. T
20. The solar radiation pressure pulls small particles toward the Sun. F

Multiple Choice

1. The solar equatorial rotation rate relative to the Earth is _____ days.
   A. 1   *D. 27.3
   B. 15   E. 59
   C. 25.4

2. The dark lines on the solar spectrum are known as _____ lines.
   *A. absorption   D. hot
   B. continuous   E. none of these
   C. missing

3. A bright-line spectrum is
   A. emitted by a luminous solid or liquid
   *B. emitted by a luminous rarefied (low-pressure) gas
   C. produced when light passes through a cool gas
   D. nonexistent
   E. real but is not produced by A, B, or C

4. The three laws governing continuous, absorption, and emission spectra are knows as _____ laws.
   A. Kepler's   *D. Kirchhoff's
   B. Newton's   E. Van Allen's
   C. Oort's

5. The proton-proton chain is an example of a(n) _____ reaction.
   *A. nuclear   D. coronal
   B. chemical   E. auroral
   C. mechanical

6. Food burning in a hot pot is an example of a(n) _____ reaction.
   A. nuclear   D. coronal
   *B. chemical   E. auroral
   C. mechanical

111

7. The opacity of the Sun is caused by
   A. positive helium ions   D. positive hydrogen ions
   B. negative helium ions   E. none of the above
   *C. negative hydrogen ions

In the figure shown below:

8. The core of the Sun is __A__.

9. The convective zone of the Sun is __B__.

10. The photosphere of the Sun is __C__.

11. The chromosphere of the Sun is __D__.

12. The corona of the Sun is __E__.

13. The prominences on the Sun are denoted by __F__.

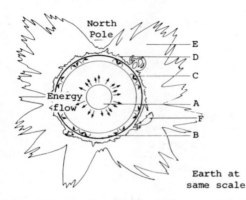

14. Throughout most of the Sun's volume, energy is transferred by
    A. conduction   D. opacity
    B. convection   E. decay
    *C. radiation

15. In the region just below the photosphere of the Sun, energy is transferred by
    A. conduction   D. opacity
    *B. convection   E. decay
    C. radiation

112

16. When one end of a metal rod is stuck into a fire, the other end gets hot also. This is an example of energy transfer by
   *A. conduction
   B. convection
   C. radiation
   D. opacity
   E. decay

17. A gas with many ions is called
   A. neutral
   B. flarish
   *C. a plasma
   D. ionic
   E. none of these

18. The sunspot cycle is approximately _____ years long.
   A. 1
   B. 5
   C. 11
   *D. 22
   E. 31

19. The solar constant is _____ in SI units (kW/m^2).
   A. 3×10^5
   *B. 1.368
   C. 1368
   D. 0.290
   E. 206,265

20. What we might term the visible surface of the Sun is the
   *A. photosphere
   B. chromosphere
   C. corona
   D. prominence
   E. plage

21. The zone of the Sun's atmosphere, just above its opaque zone, is the
   A. photosphere
   *B. chromosphere
   C. corona
   D. prominence
   E. plage

22. The outermost layer of the Sun's atmosphere is the
   A. photosphere
   B. chromosphere
   *C. corona
   D. prominence
   E. plage

23. The charged atomic particles (mostly protons and electrons) emitted nearly continuously by the Sun are called the
   A. photosphere
   *B. solar wind
   C. solar emission
   D. sunspots
   E. none of these

24. In the photosphere, there are bright regions caused by rising hot gases and darker regions of cooler falling gases. This causes the mottling called
   A. photosphere
   *B. granulation
   C. faculae
   D. sunspots
   E. corona

113

25. Dark regions on the Sun's visible surface, which may be visible to the naked eye, are called
    A. solar dark regions
    B. granulation
    *C. sunspots
    D. solar wind
    E. plages

26. Spikes of gas rising vertically through the chromosphere are
    A. sunspots
    B. flares
    *C. spicules
    D. prominences
    E. none of these

27. A sudden brightening of an area of the Sun's chromosphere, observed most commonly near sunspots, is termed a _____. This process throws material into the corona and emits vast amounts of ultraviolet rays and X-rays.
    A. spicule
    *B. flare
    C. plage
    D. granulation
    E. photosence

28. When we look toward the edge (the limb) of the Sun, we see less deeply into the photosphere. This produces what is (are) termed
    A. spicules
    B. aurorae
    C. Maunder minima
    *D. limb darkening
    E. limb brightening

29. Granulation is caused by
    *A. rising and sinking gas columns in the photosphere
    B. rising and sinking gas columns in the chromosphere
    C. rising and sinking gas columns in the corona
    D. sudden flares
    E. magnetic fields causing local heating in the chromosphere

30. The Sun has a strong magnetic field except near the sunspots, where it is weak.
    A. correct
    B. wrong: weak everywhere
    C. wrong: strong everywhere
    *D. wrong: strong near sunspots, weak everywhere else
    E. wrong: no magnetic field anywhere

31. A spectroheliograph
    A. records the solar wind
    *B. takes a picture of the Sun in a narrow spectral region
    C. produces an artificial eclipse by occulting the photosphere with a black disk
    D. measures time
    E. does none of the above

32. Astronomers have recently been unable to detect as many _____ from the Sun as expected.
   A. photons
   B. sunspots
   C. granules
   D. supergranules
   *E. neutrinos

Essay

1. What are Kirchhoff's laws? (See p. 316.)
2. What causes granulation? (See p. 323.)
3. Why is the sunspot cycle said to be 22 years instead of 11 years? (See pp. 327–328.)
4. What is the Maunder minimum? What consequences might it suggest for mankind? (See pp. 334–335.)
5. What are cosmic fuels? Contrast a global economy based on cosmic fuels with one based on fossil fuels concentrated in certain countries. (See p. 336.)

Word Practice

1. _____ laws relate emission and absorption spectra to gas density and temperature.
2. In most of the Sun's volume, energy moves primarily by _____.
3. Fire is an example of a(n) _____ reaction.
4. If energy is carried by actual matter flow, we have _____.
5. The Sun generates most of its energy by the _____ chain.
6. The atomic bomb is an example of a(n)_____ reaction.
7. Under high pressure, a hot gas emits a(n)_____ spectrum.
8. The Sun's energy is generated in the_____ _____.
9. A cool gas between a light source and an observer will cause_____lines.
10. The rusting of iron is an example of a(n) _____ reaction.
11. _____ is the study of the colors of light emitted by glowing bodies.
*12. _____ law relates the temperature to the peak color of a body.
13. The Sun's energy source is a(n) _____ reaction.
14. The discrete particle of light is the _____.
15. A(n) _____ allows a photograph of the Sun to be made in a single wavelength.
16. A hot gas under low pressure produces _____ lines.
17. _____ is the transfer of energy by adjacent molecules bumping each other.
18. A(n) _____ is the array of colors from a source in order of wavelength.

19. The _____ _____ line is a strong red hydrogen line.

20. The _____ is the visible surface of the Sun.

21. The solar _____ can be seen from the motion of sunspots.

22. Electromagnetic energy oscillations are seen as _____.

23. A(n) _____ is a gas with many ions.

24. The rate at which solar radiation reaches the Earth is called the _____ _____.

25. The solar_____ is an outrush of gas from the Sun.

26. The _____ is the Sun's lower atmosphere, just above the visible surface.

27. Solar radiation exerts a(n) _____ _____ on small dust particles and expels them from the solar system.

28. The _____ is the Sun's upper atmosphere.

29. The solar _____ is about 22 years.

30. Large blasts of material associated with sunspot sites are called _____.

31. The _____ is a vivid glow caused by solar particles entering the Earth's magnetic field.

32. A cool region on the Sun's surface controlled by magnetic fields is called a(n) _____.

33. The opacity of the photosphere is mainly caused by negative _____ ions.

34. The _____ _____ is the period from 1645 to 1715 during which sunspot numbers were very low.

35. Energy from basic planetary or astronomical sources is called _____ fuel.

36. The _____ _____ _____ consist of ions trapped in the Earth's magnetic field.

37. _____ are large clouds of gas shot out of sunspot regions.

38. _____ are convection cells in the photosphere.

39. A(n) _____ produces artificial eclipses of the Sun.

40. _____ are columns of gas between the photosphere and the chromosphere.

41. The _____ of a gas is its ability to obscure light passing through the gas.

42. A gas composed of atoms or molecules with one or more electrons free from the atom or molecules is said to be a(n) _____ gas.

43. The _____ is a nearly (or completely) massless particle emitted when two protons fuse to produce 2H.

Answers to Word Practice

1. Kirchhoff's
2. radiation
3. chemical
4. convection
5. proton-proton
6. nuclear
7. continuous
8. solar core
9. absorption
10. chemical
11. spectroscopy
12. Wien's
13. nuclear
14. photon
15. spectroheliograph
16. emission
17. conduction
18. spectrum
19. hydrogen alpha
20. photosphere
21. rotation
22. light (or color)
23. plasma
24. solar constant
25. wind
26. chromosphere
27. radiation pressure
28. corona
29. cycle
30. flares
31. aurora
32. sunspot
33. hydrogen
34. Maunder minimum
35. cosmic
36. Van Allen belts
37. prominences
38. granules
39. coronograph
40. spicules
41. opacity
42. ionized
43. neutrino

Chapter 16
Measuring the Basic Properties of Stars

Answers to Problems in the Text

1. Because most stars are much smaller in angular size (the Sun at one parsec would have an angular size of $0.''009$), we could resolve only the closest large stars, such as Betelgeuse. But the Earth's atmosphere often degrades the useful resolution to $0.''5$ or $1''$. So a ground-based telescope is often useless for stellar diameters. On the Moon, there would be no atmosphere to blur the image (and no scattered light either). At higher magnifications, atmospheric movement and blur are magnified and the images actually look worse, as well as fainter.

2. 1 parsec = 3.26 light-years; 20 parsecs = 65.2 light-years. So, 65.2 years.

3. Hydrogen (80%) and helium (20%) and some heavy elements.

4. The bluish is hotter (Wien's law): Color is temperature. The bluish star has higher luminosity: Of two stars with the same surface area, the hotter star emits more per unit area.

   $E = \sigma T^4 A$ (Stefan-Boltzmann law)

5. Reddish, by the reasoning in problem 4: $E = \sigma T^4 A$. T is lower for reddish stars, so area A must be larger for the star to radiate the same energy per second (E).

6. It is nonlinear. Doubling the amount of light (brightness) does not double magnitude. In fact, a hundredfold increase only decreases magnitude by an additional five units. Also, as the quantity measured gets larger (that is, brightness increases), the magnitude gets smaller.

7. It has the same temperature (as judged from the spectrum). It is not moving radially with respect to the Sun (that is, it has no Doppler shift). It is probably the same size (again, the spectrum is the same). It must be far away from us; it is faint. 30 magnitude $\approx 10^{12}$ times fainter, so the star must be $\sqrt{10^{12}} = 10^6$ times as far away as the Sun. Distance = 10^6 AU or about 5 parsecs. The star must be very similar in chemical composition to the Sun (again, judging from the spectrum).

8. Distance in parsecs = 1/(parallax in seconds of arc), so
   $$\text{distance} = \frac{1}{5 \times 10^{-2}} = \frac{100}{5} = 20 \text{ parsecs.}$$

9. If the stars are equidistant from the center of mass, they have equal mass. The sum of the masses is 5 solar masses, so mass = 2.5 $M_\odot$.

10. The surface temperature of the Sun is such as to particularly excite calcium rather than hydrogen. The strength of lines is influenced mainly by gas conditions, not gas composition.

11. Rapid rotation or a turbulent atmosphere, causing both red and blue Doppler shifts.

12. Difference in magnitude = 26.5 − 12.5 = 14. Each 5 = 100 factor; each 1 = 2.5 factor.

$$100 \times 100 \times 2.5 \times 2.5 \times 2.5 \times 2.5 = \frac{100 \times 100 \times 100}{2.5}$$

$$= \frac{10}{2.5} \times 10^5$$

= 4 × 10⁵ times as bright. An exposure of the Moon must be 4×10^5 times as long as an exposure of the Sun. Thus, a $\frac{1}{100}$-s exposure must be:

$$\text{Time} = \frac{1}{100} \times 4 \times 10^5 \text{ s} = 4 \times 10^3 \text{ s} \sim 1 \text{ hour.}$$

13. a. The red shift means it is moving away.
    b. Its speed is 1% of the speed of light = 300 km/s, so it is moving unusually fast.

14. In one year, the star moves an angle of 1" so the distance traveled per year is given by the small angle equation

$$\frac{1"}{2.1 \times 10^5} = \frac{d}{D}, \text{ where } D = 10 \text{ pc} = 3.08 \times 10^{16} \text{ m} \times 10;$$

so $d = \dfrac{1 \times 3.08 \times 10^{17}}{2.1 \times 10^5}$ m = 1.5×10^{12} m = 1.5×10^9 km. Divide this by the

number of seconds per year ($\sim \pi \times 10^7$) to set the velocity in km/s:

velocity = 49 km/s.

15. The star is receding at 300 km/s and moving at right angles to the line of sight at 49 km/s. The space velocity is the long side of the triangle whose other sides are the radial and tangential velocities, so

$$V_{\text{space}} = \sqrt{V^2_{\text{radial}} + V^2_{\text{tangential}}} = 304 \text{ km/s.}$$

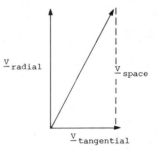

16. Use the Stefan-Boltzman law (Optional Equation VIII):

$$\Sigma = \sigma A T^4$$

The problem is easier if we do a ratio. Let $*$ stand for the star and $\odot$ for the Sun.

$$\frac{\Sigma_*}{\Sigma_\odot} = \frac{\sigma A_* \, T_*^4}{\sigma A_\odot \, T_\odot^4} = \frac{A_*}{A_\odot} \left(\frac{T_*}{T_\odot}\right)^4$$

We are given $\Sigma_* = \Sigma_\odot$, so

$$A_* = A_\odot \left(\frac{T_\odot}{T_*}\right)^4 = A_\odot \left(\frac{5700}{2850}\right)^4 = 16 A_\odot.$$

Now area goes as radius squared, so
$$R* = 4R_\odot.$$

17. a. The average kinetic energy = $(3/2)kT$, so as T increases, the kinetic energy increases on the average.

b. Since the kinetic energy increases with increasing temperature, the force of collision between atoms must increase also. Thus, a collision is more likely to cause ionization or excitation with higher temperature.

c. To ionize, we need 2×10^{-18} joules. Find the temperature by $E = (3/2)kT$. That is,

$$
\begin{aligned}
T &= (2/3)E/k \\
&= (2/3) \times 2 \times 10^{-18} \text{ J}/1.38 \times 10^{-23} \text{ J/K} = 9.7 \times 10^4 \text{ K}.
\end{aligned}
$$

We could argue that each atom needs only half the ionization energy so that one needs only

$$T = 48,000 \text{ K}.$$

d. In any event, this is a ballpark figure near the 20,000–40,000 K temperature of O and B stars.

Sample Test Questions

True-False

1. Most of the brighter stars have English names. F
2. The disks of stars seen through the telescope are the images of the stellar diameters. F
3. The parsec is the distance light travels in one year. F
4. The light-year is a unit of time. F
5. The brightness of a star detected by an observer on Earth is termed apparent brightness. T
6. The magnitude system was devised by Hipparchus about 130 B.C. T
7. The faintest star visible to the naked eye on a clear night has a magnitude of about 5. F
8. The device used to produce a photographic image of a spectrum is a spectrograph. T
9. A spectrometer produces a spectrum in the form of a chart or graph. T
10. Emission lines appear as dark vertical lines on a photographic spectrum and as notches or valleys on a scan. F
11. An M star is cooler than an A star. T
12. The stable arrangement of an atom is called its ground state. T
13. When electrons move from a higher to a lower energy level, they produce an absorption spectrum. F
14. The visible set of lines produced by hydrogen transitions ending on the $n = 2$ level is called the Balmer series. T
15. Temperature is a measure of the average speeds of atoms or molecules in a gas. T
16. If a light source is receding from us, we say it has a red shift. This means the light source will appear red. F
17. The Stefan-Boltzmann law tells us the wavelength at which most radiation from a star will be emitted. F
18. The distance limit for reliable parallaxes is about 20 parsecs. T

120

19. The luminosity of a star is the total amount of energy it radiates each second. T
20. Circumstellar material is detected from spectra by using the Zeeman effect. F
21. Star diameters are measured by determining the luminosity and temperature and then using the Stefan-Boltzmann law. T
22. Stars with nearly identical spectra usually have nearly identical mass. T
23. Stars with rapidly rising and sinking gases in convection cells will show strong rotational line broadening. F
24. The Doppler shift enables us to detect the tangential velocity of a star. F
25. The tangential velocity of a star is its motion perpendicular to the line of sight. T

Multiple Choice

1. _____ causes even perfect lenses or mirrors to form images of stars as disks surrounded by rings.
   A. the finite speed of light
   B. color
   C. the *F* -ratio
   *D. diffraction
   E. none of these

Consider the following photograph of an overexposed star image:

2. The cross is caused by
   A. diffraction on the edges of the round telescope
   B. diffraction from an obviously square telescope
   C. refraction off the secondary mirror's supporting structure
   *D. diffraction off the secondary mirror's supporting structure
   E. none of these

Referring to the same photograph:

3. The round disk is caused by
   *A. diffraction on the edges of the round telescope
   B. diffraction from an obviously square telescope
   C. refraction off the secondary mirror's supporting structure
   D. diffraction off the secondary mirror's supporting structure
   E. none of these

4. 1 parsec = _____ light-years.
   A. 2
   *B. 3.26
   C. 32.6
   D. 206,265
   E. 3×10^5

5. If two stars differ in magnitude by 5, the ratio of the amount of light received (brighter/fainter) is about _____ to 1.
   A. 1
   B. 2.5
   *C. 100
   D. 4
   E. 10,000

6. If star X appears to be 2.5 times as bright as star Z, then the two stars differ in magnitude by
   A. 0
   *B. 1
   C. 2.5
   D. 0.5
   E. 10

7. If two stars are emitting the same amount of light, the closer star will appear to be
   A. dimmer
   *B. brighter
   C. the same brightness as the farther
   D. red
   E. blue

8. Consider three stars of apparent magnitudes 5, 2, and -1. Rank them in order of increasing apparent brightness.
   *A. 5, 2, -1
   B. -1, 2, 5
   C. 2, 5, -1
   D. 5, -1, 2
   E. -1, 5, 2

9. Consider three stars of apparent magnitudes 0, 7, and 3. Rank them in order of decreasing apparent brightness.
   A. 0, 7, 3
   B. 7, 3, 0
   C. 3, 0, 7
   *D. 0, 3, 7
   E. 7, 0, 3

10. The absolute magnitude is the magnitude a star would have if it were _____ parsecs away.
    A. 1
    B. 2.5
    C. 5
    *D. 10
    E. 100

11. The Sun has an apparent magnitude of
    A. 26.5
    B. 5
    C. 0
    D. -12.5
    *E. -26.5

12. The Sun has an absolute magnitude of about
    A. 26.5   D. -12.5
    *B. 5   E. -26.5
    C. 0

Refer to the spectrum graph shown above to answer questions 13–15.

13. The continuous spectrum is __A__ .

14. An emission line is __C__ .

15. An absorption line is __B__ .

16. The letters classifying the spectral sequence of stars from red to blue (low temperature to high temperature) are
    A. BBROYGBYGW   *D. MKGFABO
    B. NRAWDOP   E. OBAFKMG
    C. OBAFGKM

17. An A5 star is
    A. halfway from B0 to A0   D. red
    B. halfway from A0 to B0   E. none of these
    *C. halfway from A0 to F0

18. M stars are _____ in color.
    A. blue   D. orange to red
    B. blue to white   *E. red
    C. white to yellow

19. Two rapidly moving atoms collide, exciting an electron in one atom to a higher energy level. Later this electron returns to its original energy level. This latter process produces a(n) _____ spectrum.
    *A. emission   D. offline
    B. absorption   E. none of these
    C. desorption

123

20. G stars have prominent spectral lines of
    A. ionized metals
    B. hydrogen Balmer
    C. helium
    *D. both A and B
    E. all of these, A, B, and C

21. The Bohr model of the atom puts the _____ in shells around the nucleus.
    A. protons
    B. neutrons
    *C. electrons
    D. partons
    E. positrons

22. An atom is said to be excited when
    A. one or more electrons are removed
    *B. one or more electrons are in higher energy levels than normal
    C. one or more protons are in higher energy levels than normal
    D. one or more protons are removed
    E. it is in a gas

23. An atom is said to be ionized when
    *A. one or more electrons are removed
    B. one or more electrons are in higher energy levels than normal
    C. one or more protons are in higher energy levels than normal
    D. one or more protons are removed
    E. it is in a gas

24. The change in observed wavelengths from a wave source that is approaching or receding from us is the _____ effect.
    A. Michelson
    B. Roemer
    C. Einstein
    D. Hartmann
    *E. Doppler

*25. If a star is approaching us at 0.01 times the speed of light and it is emitting 500 nm of light, we will see a wavelength shift of
    A. 5 nm to the red
    B. 0.5 nm to the red
    C. 0.05 nm to the red
    *D. 5 nm to the blue
    E. no shift

*26. A star emitting light at a wavelength of 4.50×10^{-7} m is receding from us at 0.1 times the speed of light. We will see the light at a wavelength of
    _____ m.
    A. 4.05×10^{-7}
    *B. 4.95×10^{-7}
    C. 1×10^{-4}
    D. 6.8×10^{-7}
    E. 4.50×10^{-7}

27. If the temperature of a perfect radiator is doubled, the energy emitted per unit area per unit time will be _____ times as large as originally.

A. $\frac{1}{2}$ D. 2
B. 8 *E. 16
C. 4

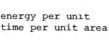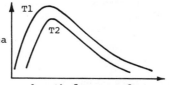

energy per unit
time per unit area

wavelength for a perfect radiator

28. Refer to the diagram above. The perfect radiator with the greater temperature is
   *A. T1 C. both are equal
   B. T2 D. not enough information

*29. Which would emit more energy: star 1 of area 16A and temperature T or star 2 of area A and temperature $2T$?
   A. star 1 *C. both would be the same
   B. star 2 D. not enough information

*30. The Sun has a diameter of 1.4×10^{11} km and a surface area of 6.1×10^{19} m^2. If the temperature of the Sun is 5700 K and the constant in the Stefan-Boltzmann law is 5.7×10^{-8} W/m^2 • K^4, what is the approximate energy per unit time that the Sun emits in watts?
   A. 4.0×10^{20} D. 3×10^{10}
   *B. 4.0×10^{33} E. 4.0×10^{15}
   C. 4.0×10^{45}

31. The parallax of a star is the angle through which the star appears to move (against the background stars) in the course of a year (corrected for its relative position to the ecliptic).
   A. correct
   B. wrong: It is one-fourth the angle in a year.
   *C. wrong: It is half the angle in a year.
   D. wrong: The parallax of a star is its temperature divided by one-half its mass.
   E. wrong: Parallax has to do with color changes.

32. A possible definition for the parsec is the distance to a star whose parallax is _____ seconds of arc (corrected for its relative position to the ecliptic).
   A. 0.1 D. 0.5
   *B. 1.0 E. none of these
   C. 10

33. If a star has a parallax of 1 second of arc, its distance is _____ parsecs.
    A. 2   D. 10
    *B. 1   E. 13
    C. $\frac{1}{2}$

34. If a star has a parallax of 0.1 second of arc, its distance is
    _____ parsecs.
    A. 2   *D. 10
    B. 1   E. 13
    C. $\frac{1}{2}$

35. The distance limit for reliable parallaxes is _____ parsecs.
    A. 1   *D. 20
    B. 5   E. 50
    C. 10

36. Which stars must have a large positive luminosity?
    *A. intrinsically bright stars
    B. intrinsically dim stars
    C. red stars
    D. no stars
    E. both B and C

37. Which stars must have large positive absolute magnitude?
    A. intrinsically bright stars
    *B. intrinsically dim stars
    C. red stars
    D. no stars
    E. both B and C

38. The magnitude based on all of the electromagnetic energy ideally reaching the Earth
    is
    A. visual magnitude
    B. photographic magnitude
    C. photovisual magnitude
    *D. bolometric magnitude
    E. none of these

39. The bolometric magnitude is useful because it
    *A. takes into account radiation at all wavelengths
    B. corrects for year of observation
    C. is the easiest to measure
    D. is the same for all stars
    E. is always a positive number

40. The rapid motion of atoms in a star's photosphere causes
   A. rotation
   B. photons
   C. collisional broadening
   *D. Doppler broadening
   E. none of these

41. When we look at the center of the disk of the Sun (or any star), we see deeper into the photosphere than when we look at the limb. This causes
   A. collisional broadening
   *B. limb darkening
   C. limb brightening
   D. limb decoloration
   E. none of these

42. We can get the temperature for a star from
   A. measuring its continuous spectral distribution
   B. studying its line spectrum
   C. looking at the total energy per unit area the star emits
   D. both A and B
   *E. all three, A, B, and C

43. We can measure the radial velocity of a star from the
   *A. Doppler shifts of its spectral lines
   B. Doppler broadening of its spectral lines
   C. presence or absence of particular lines
   D. total energy per unit area emitted
   E. none of the above

44. We can get the chemical composition of a star by studying its
   A. mass
   B. Zeeman effect
   *C. spectrum
   D. opacity
   E. Doppler shifts

45. We can get the magnetic fields on a star's surface by studying its
   A. mass
   *B. Zeeman effect
   C. spectra
   D. opacity
   E. Doppler shifts

46. Which would be a normal unit with which to express proper motion?
   A. light-years per century
   *B. seconds of arc per year
   C. centimeters per second of time
   D. kilometers per year
   E. furlongs per fortnight

Essay

1. Distinguish between apparent and absolute magnitude. (See pp. 341–343.)
2. Compare some typical apparent magnitudes. (See Table 16-1.)
3. List the seven main spectral classes and the typical color associated with each. (See Table 16-2.)

4. What are the two important spectroscopy laws of this chapter? (See pp. 347–351.)
5. List eight of the 12 important stellar properties we studied and the techniques for measurement. (See Table 16-3.)
6. Why is spectroscopy so important for astronomy? (See pp. 346–347.)

Word Practice

1. The _____ system is a scale for measuring brightness of stars and other astronomical objects.

2. A(n) _____ _____ produces an emission line.

3. The _____ _____ of stars are arranged by temperature, as shown by the spectral lines.

4. The _____ magnitude of a star is its magnitude as we perceive it from the Earth.

5. The _____ _____ is how bright the stars would appear if they were placed at some identical, standard distance.

6. A(n) _____ produces a spectrum in the form of a chart or graph.

*7. An atom is _____ if one or more electrons are in higher energy levels than usual.

8. A(n) _____ is the distance light travels in one year.

9. The _____ magnitude of a star is the magnitude the star would have if it were 10 pc from us.

10. The spectral classification scheme is a(n) _____ classification.

11. A(n) _____ produces a spectrum, which is a photographic image.

12. The first letter of the Greek alphabet is _____.

13. The _____ alpha line is a prominent red line in the hydrogen spectrum.

14. The _____ is the general level of brightness between absorption or emission lines.

15. A(n) _____ is the distance at which 1 AU appears to be 1 second of arc in angular size.

16. A(n) _____ consists of a nucleus with a surrounding cloud of electrons.

17. The _____ disk is produced by diffraction.

18. A hot, low-pressure gas produces _____ lines.

19. The _____ _____ of a star is how bright the star appears from Earth.

20. Electronic orbits in atoms are also known as _____ _____.

21. _____ causes a star image to appear as a disk surrounded by faint rings.

*22. The Bohr model of the hydrogen atom has the electrons in _____ _____.

23. The _____ series is a set of visible lines of hydrogen ending on the $n = 2$ level.

24. _____ are objects that generate (or have generated) energy in their cores by nuclear reactions.

25. The _____ _____ magnitude of a star is measured using a system with the same color sensitivity as the human eye.

26. _____ is a measure of the average velocities of atoms or molecules in a gas.

27. Dark vertical lines on a photographic spectrum and notches or valleys on a scan are _____ lines.

28. _____ is the study of spectra.

29. A(n) _____ is the distribution of light into different wavelengths.

30. The _____ equation tells us the relative numbers of atoms in each state of excitation.

Answers to Word Practice

1. magnitude	16. atom
2. ionized atom	17. Airy
3. spectral classes	18. emission
4. apparent	19. apparent brightness
5. absolute brightness	20. energy levels
6. spectrometer	21. diffraction
7. excited	22. quantized orbits (or energy levels)
8. light-year	23. hydrogen (or Balmer)
9. absolute	24. stars
10. temperature	25. visual apparent
11. spectrograph	26. temperature
12. alpha	27. absorption
13. Balmer (or hydrogen)	28. spectroscopy
14. continuum	29. spectrum
15. parsec	30. Saha

Chapter 17
The Systematics of Nearby Stars: The H-R Diagram

Answers to Problems in the Text

1. a. Main-sequence stars, spectral classes, G-M, masses usually less than the Sun's mass.
   b. About 20% or less. (Use Table 7-1; assume unknown masses $< M_\odot$.)
   c. No. Most apparently bright stars are giants. Giant stars are rare.
2. No. They are larger in diameter. $E = \sigma T^4 A$. Thus, if they are of the same spectral class as the Sun, the T s are equal. Thus, for a giant, denoted by $*$, and the Sun, indicated by $\odot$,

   $$E_*/E_\odot = L_*/L_\odot = A_*/A_\odot = 100. \quad \text{(See Fig. 17-8.)}$$

   Thus, surface area is a factor of 100 larger, which means the diameter of the giant is a factor of 10 larger than the diameter of the Sun. This result checks with Fig. 17-5.
3. a. upper and lower left-hand corners
   b. upper right-hand and left-hand
   c. upper right-hand
   d. upper left-hand

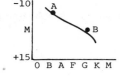

4. a. usually the one to the upper left
   b. the one toward the upper left
   c. same
   d. same
   Thus, star A in all cases.
5. The more massive will evolve faster, reaching main sequence first, becoming a giant star first, and also revolving to a white dwarf first. It is also more likely to become a supernova.
6. In 5–6 billion years, the Sun will rapidly grow (within a period of a few million years) to a red giant about 100 times its present size. From the Earth, the Sun would appear very roughly $100 \times 1/2 \sim 50°$ in diameter. Mercury and Venus (probably) would be absorbed by the Sun. The Sun would be cooler but more luminous (emitting more energy/second), so the Earth would heat and the oceans would evaporate. Life forms would perish from the heat. As one astronomer has put it, "Conditions would be quite miserable." Then, in just a few million more years, the Sun would shrink to a white dwarf. Perhaps it would be a variable star along the way. Everything on the Earth would freeze, assuming the planet had not been engulfed in the Sun's outer part during the red giant stage.
7. Temperature and luminosity are fixed by the two coordinates of the H-R diagram. These fix the radius. Only the mass is not fixed by position on the H-R diagram, unless the star is known to be on the main sequence. For example, stars of different mass may be in the giant region.
8. Rate of concentration slows. Objects much less than 0.08 of the Sun's mass never become stars because they do not get hot enough inside to start nuclear reactions. Thus, Jupiter is not a star.

9. Main-sequence life for a sunlike star is 9 billion years. Our estimates are that the solar system is 4.6 billion years old and took about 100 million years to form. This is consistent with data. The Sun has been roughly the same for the last 4 billion years. The solar system could hardly be 4.6 billion years if the Sun were only 1 billion years old.

10. Radiation from the Sun's center exerts a pressure that supports the star. This pressure lasts as long as nuclear reactions create energy. It takes a long time to burn up the hydrogen in normal stars.

11. There is now more helium and less hydrogen and also some additional heavier elements. The only difference is in the core, where reactions have taken place. There is no mixing with most of the Sun's volume.

12. Blanket statement not correct. Teachers should distinguish "chronological age" (years) from "evolutionary age" (stage). A massive star evolves faster than a low-mass star. Many of the chronologically very old $1/10\ M_\odot$ stars have yet to reach the main sequence. The Sun is chronologically older than a $10\ M_\odot$ red giant star.

Advanced Problems

13. $W\ =\ 0.0029/T$
    $W_M\ =\ 0.0029/2900\ \text{K} = 10^{-6}\ \text{m} = 1000\ \text{nm}$, greater than visible wavelength: infrared.
    $W_K\ =\ 0.0029/29,000\ \text{K} = 10^{-7}\ \text{m} = 100\ \text{nm}$, shorter than visible light: ultraviolet light.

    Each star must show color characteristic of the end of the spectrum at which it lies, so an M star is very red and a K star is very blue. Both peaks lie outside the visible range.

14. a. Energy radiated is proportional to T^4 by Stefan-Boltzmann law: $(20,000/2000)^4 = 10^4$.
    b. Use line $10\ R_\odot$, and read off the luminosity difference on the left-hand scale. From 2500 K to 20,000 K, it looks like about 10,000 or 20,000 in luminosity. This checks approximately.

15. a. Density $= \dfrac{\text{mass}}{\text{volume}}$. For this star, mass $= 0.8\ M_\odot$; volume $= \dfrac{4}{3}\pi R^3 = \dfrac{4}{3}\pi(0.015R_\odot)^3$; density $= 2.4 \times 10^5 \times$ density of Sun. Using Table 8-1, we find density of Sun $= 1.4 \times 10^3\ \text{kg/m}^3$, so density $= 3.4 \times 10^8\ \text{kg/m}^3$. This is between 10^4 and 10^5 times the density of normal matter.
    b. Assume the question means the Earth. We need to calculate the volume of a matchbox and multiply by acceleration due to gravity and by density. Weight $=$ density $\times$ volume $\times$ 9.8 m/s^2. This gives weight in newtons.

131

Sample Test Questions

True-False

1. The principal differences in the stellar spectra of main-sequence stars arise in general from variations in a single condition in the stellar atmosphere. T
2. The letters in the H-R diagram stand for Herman-Russell. F
3. Unusually small white stars are called white dwarfs. T
4. On the H-R diagram, a line of constant star radius slants down from left to right. T
5. The stars with an apparent magnitude brighter than first magnitude are called representative stars. F
6. The 40 known stars within 4 parsecs of the Sun are known as representative stars. T
7. There are more red giants and supergiants on the list of prominent stars than are found on the list of representative stars. T
8. Stars stay on the main sequence as long as the internal chemistry and energy production change rapidly. F
9. On the main sequence, a star with less mass than the Sun's will be brighter than the Sun. F
10. The relationship between mass and magnitude, called the mass-luminosity relation, holds for all stars. F
11. The equilibrium structure of an ordinary star is determined uniquely by its mass and past history. F
12. The Russell-Vogt theorem explains the cause of stellar evolution. T
13. Stars evolve through most of their life cycle with essentially constant composition but changing mass. F
14. As the Sun evolves, it will first become cooler and brighter. T
15. The more massive a star, the more rapidly it evolves. T
16. The H-R diagram can serve as a tool for dating groups of stars that formed together. T

Multiple Choice

Refer to Diagram 1 for questions 1–4.

1. The spectral class of star A is
   A. G5
   B. A3
   C. M0
   *D. B7
   E. M6

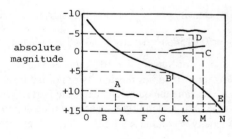

Diagram 1

132

2. The spectral class of
   star C is
   A. G5
   *B. M0
   C. M6
   D. A3
   E. B7

3. The star that has an absolute magnitude of +10 is ___A___.

4. The star that is a supergiant is ___D___.

Refer to Diagram 2 for questions 5–9.

5. The A0 white dwarf
   star is ___A___.

6. The absolute magnitude
   of the A0 white dwarf
   star is
   A. +4
   B. +6
   *C. +11
   D. -1
   E. -7

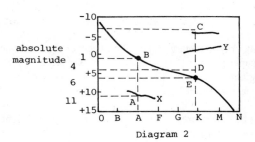

Diagram 2

7. The spectral class of
   the star with absolute
   magnitude -7 is
   A. A0
   *B. G8
   C. K2
   D. B3
   E. Z2

8. The supergiant is ___C___.

9. Of the stars on Diagram 2 labeled A–E, the brightest of the hotter stars is
   ___B___.

10. In Diagram 2, of the stars in the main sequence, the most massive is ___B___.

Refer to Diagram 3 for
questions 11–12.

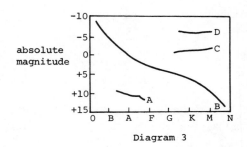

Diagram 3

11. Main-sequence stars
    are denoted by line
    __B__.

12. Supergiant stars
    are denoted by line
    __D__.

13. Consider the following data on five stars. The hottest star is

Star	Apparent Magnitude	Spectrum
A	15	G2 main sequence
B	20	M3 supergiant
C	10	M3 main sequence
*D	15	B9 main sequence
E	15	M5 main sequence

14. Small, hot, faint stars, often called degenerate, are termed
    A. supergiants   *D. white dwarfs
    B. giants   E. none of these
    C. main sequence

15. Most of the apparently bright stars in the sky are bright because they are close to our
    Sun (within 4 parsecs).
    A. correct
    *B. wrong: Most are intrinsically bright and lie far from the Sun.
    C. wrong: Close means closer than 1 parsec.
    D. wrong: This is true only for the green ones.
    E. wrong: Both C and D changes are needed.

16. The Russell-Vogt theorem tells us that the equilibrium structure of an ordinary star is
    determined uniquely by its
    A. mass   *D. both A and C
    B. age   E. all three, A, B, and C
    C. chemical composition

17. The most common star masses are _____ $M_\odot$.
    A. 0.001–0.01   D. 1.0–10
    B. 0.01–0.1   E. 10–100
    *C. 0.1–1.0

18. Refer to Diagram 4. Which line shows the approach to the main sequence?
    A. line I   D. line IV
    B. line II   E. none of these
    *C. line III

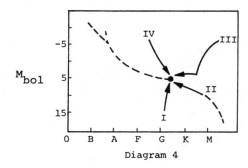

Diagram 4

19. Refer to Diagram 5. Suppose a star is located at X on the main sequence. Which direction will it move as it evolves off the main sequence?
    A. direction A   D. direction D
    B. direction B   E. none of these
  *C. direction C

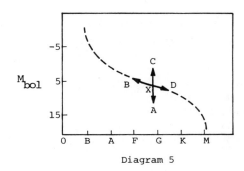

Diagram 5

20. Refer to Diagram 6. Which part of the path is the evolution from the red giant stage toward the white dwarf stage?
    A. 6 to 5
    B. 1 to 2
    C. 5 to 4
  *D. 3 to 5
    E. 5 to 3

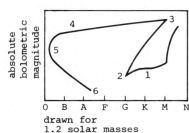

drawn for
1.2 solar masses
Diagram 6

21. Refer to Diagram 7, a Hertzsprung-Russell diagram for a hypothetical cluster from birth to old age. Place the diagrams in the correct sequence of increasing time.

A. 1, 2, 3
B. 2, 3, 1
C. 1, 3, 2
*D. 2, 1, 3
E. none of these

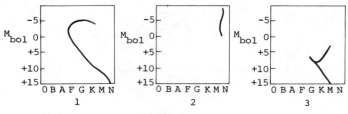

Diagram 7

22. In simple models, just before reaching the main sequence, a star has _____ magnitude (or luminosity).

*A. approximately constant
B. decreasing
C. increasing

23. The main sequence for a star can best be described as the first state in which the energy produced by _____ balances the energy lost by radiation from the surface.

A. gravitational contraction
*B. fusion of hydrogen into helium
C. fission of helium into hydrogen
D. fission of helium into carbon
E. none of the above

Essay

1. List the spectral classes in order of increasing temperature. Explain why the stars were first listed in alphabetical order but are now listed by classes in a very different order. (Answer to first part: MKGFABO.) As astronomers studied stars, they found that the original order was wrong and that some classes did not really exist.

2. Distinguish between prominent and representative stars. (See pp. 366–368.)

3. If there are a giant star, a supergiant star, and a main-sequence star all of the same spectral class, say K0, what can you say about their relative sizes? Why? (Size: largest, supergiant; next, giant; smallest, main sequence. Temperature: same. Definition of spectral classes and Stefan-Boltzmann law.)

4. What is the Russell-Vogt theorem? (See pp. 374–376.)

5. Draw an H-R diagram, label the axes, and indicate main-sequence, giant, supergiant, and white dwarf regions. (See Diagram 3 in Sample Multiple Choice Questions: A, white dwarf; B, main-sequence; C, giant; D, supergiant.)

6. Explain why the H-R diagram has been useful in interpreting and discussing star data and theory, both historically and in the present. Give examples of how it is used in such discussions.

136

(Main points: It can be plotted with observational data such as *M* versus spectrum [as Hertzsprung and Russell did] or with theoretical data [such as temperature and luminosity]. It can be used for describing clusters of stars, showing dwarfs and giants.)

Word Practice

1. Stars undergo change as they age. This is called stellar _____.

2. Most stars on the H-R diagram are _____ _____.

*3. Stars in the upper half of the H-R diagram are intrinsically _____.

4. The _____ theorem states that "the equilibrium structure of an ordinary star is determined uniquely by its mass and chemical composition.

5. Stars with an apparent magnitude less than 1 (these are bright stars) are called _____ stars.

*6. One of the astronomers who first plotted absolute magnitude against spectral class was _____.

7. A plot of absolute magnitude against spectral class is the _____ diagram.

8. _____ _____ are small hot stars.

*9. Stars to the right on the H-R diagram are _____.

10. The path a star traces out on the H-R diagram as it evolves is the star's _____ _____.

*11. Stars on the lower half of the H-R diagram are intrinsically _____.

12. A statement of the _____ relation is that "on the main sequence each different mass corresponds to a different luminosity."

13. Stars that suddenly and unpredictably brighten are called _____.

14. Large red stars are _____ stars.

15. The stars within 4 pc of the Sun are called _____ stars.

16. The largest of the red stars are the _____ stars.

17. Stars to the left on the H-R diagram are _____.

*18. A tabulation (or display) of the relative amounts of light received from a star as a function of the wavelength of the light is the star's _____.

1. evolution
2. main sequence
3. bright
4. Russell-Vogt
5. prominent
6. Hertzsprung or Russell
7. spectrum-luminosity
8. white dwarfs
9. cool
10. evolutionary track
11. dim
12. mass-luminosity
13. novae
14. giant
15. representative
16. supergiant
17. hot
18. spectrum

Chapter 18
Stellar Evolution I: Birth and Middle Age

Answers to Problems in the Text

1. The last half.
2. a. Hinder, because there would be no areas of greater density.
   b. Any substantial enough random collection of material will begin to collapse under gravitational attraction. Supernovae and novae shear from galactic rotation; passage of stars through the gas might cause dense regions.
   c. Yes.
3. Fragmentation is an almost inevitable consequence of the cloud collapse process. Because each large cloud that has many star masses breaks up into subclouds that form stars, many stars form together.
4. The two time scales are consistent. Star formation theories give us stars forming inside cocoon nebulae, which would also provide the material for planetary formation. And, indeed, silicate dust has been observed to be near young stars. For planetary theory, we have the data from the solar system and satellite system, plus samples from the Moon and meteorites. Much of the data we have collected is geochemical and dynamical. For star formation, we have only what we can observe by telescope in a time that is short compared to the lifetimes of stars.
5. Even these short evolutionary time scales are long compared with the length of time in which we have been observing the stars. Of the few changes that have occurred, most are too small for us to be sure what consequences they imply.
6. The very high-mass end would have evolved off the main sequence because those stars evolve rapidly. The low-mass stars would not yet have reached the main sequence.
7. We would not see much with the naked eye, but infrared telescopes would see an object that was brightening and shrinking in size. The star would reach its high-luminosity phase in about 1000 years and then spend several million years fading toward the main sequence. During this time, the star would probably be hidden by its cocoon nebula, so it would be visible only in infrared. But after the first thousand years, it would be hard to detect even by infrared observation.
8. Many clouds are too low in density and size to collapse. Even some dense clouds have enough turbulence to keep gravity from causing them to collapse.
9. At that point the internal temperature is high enough for hydrogen to fuse into helium, releasing energy. This energy provides a pressure that prevents the collapse. A star with very large amounts of hydrogen in its core will have a long period of stability.

10. W = 10 μm = 10^{-5} m

    T = 0.00290/W = 290 K

    v/c = $\dfrac{\text{change in wavelength}}{\text{wavelength}}$ = $10^{-3} \rightarrow v = 10^{-3} c$

    = 300 km/s

We have a G-type star surrounded by an opaque dust cloud. The dust cloud has been heated to 290 K and probably contains silicates (other solids might be present as well). The blue shift indicates that the part of the dust cloud between us and the star is approaching us at a speed of 10^{-3} times the speed of light = 300 km/s. Two interpretations:

a. Between us and the G-type star is a warm dust cloud of T = 290 K approaching us. This is unlikely since the cloud is too warm to be truly interstellar; it must be heated by being near a star.

b. A newly born G-type star is in the process of blowing away its cocoon nebula. Thus, it is very much like the hypothesized early solar system.

Explanation b is the most probable because the interstellar cloud would not have a heat source.

Sample Test Questions

True-False

1. We have only theory, not proofs, that stars are being born today. F
2. We know of massive stars that must be only a few million years old. T
3. On the average, in space there are only 1–5 atoms/cm^3. T
4. In star-forming regions, there may be as many as 10,000 atoms and molecules/cm^3. T
5. The relationship needed between mass and density for a gas cloud to collapse is given by the Jeans theory. T
6. A collapsing gas cloud first subdivides into regions containing enough mass for thousands of stars. T
7. The average density of the universe is believed to be 10^{-27} or 10^{-28} kg/m^3. T
8. Hayashi was one of the astrophysicists who pioneered in calculating evolutionary tracks. T
9. The Hayashi phase is a short, low-luminosity phase. F
10. Newly forming stars must be above and to the right of the main sequence. T
11. The cocoon nebula hides the star from view during its earliest formation. T
12. The cocoon nebula is eventually completely absorbed by the star it surrounds. T
13. Objects too big to be planets but too small to become stars are called brown dwarfs. T
14. Stars of sunlike mass produce their energy by the set of nuclear reactions called the carbon cycle. F
15. Dark clouds 1000 to 100,000 AU, common near young star clusters, are called T Tauri globules. F

Multiple Choice

1. Some (or one) of the proofs that stars are forming today are (is)
   A. short-lived massive stars
   B. the youth of the solar system
   C. existing dust clouds
   *D. both A and B
   E. all three, A, B, and C

2. A well-tested body of ideas that has a mathematical formulation and, usually, has been confirmed by observation is called a(n)
   A. hypothesis
   B. speculation
   C. idea
   *D. theory
   E. none of the above

3. The line on the graph that best represents the prediction of the virial theorem (Jeans theorem) is __A__.

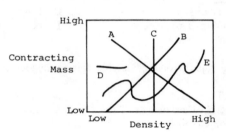

4. The energy radiated during the initial phases of stellar formation comes from
   A. resistance loss   D. nuclear fusion
   *B. Helmholtz contraction   E. nuclear fission
   C. Hayashi transport

5. The short, high-luminosity phase of stellar formation occurs when the star is below and to the left of the main sequence.
   A. correct
   B. wrong: below and to the right
   C. wrong: above and to the left
   *D. wrong: above and to the right
   E. wrong: occurs on the main sequence

6. A star of solar mass spends about a thousand years in the high-luminosity stage and reaches the main sequence after _____ years.
   A. 100   *D. 100 million
   B. 1 million   E. 10 billion
   C. 10 million

141

7. The high-luminosity phase is easily visible to an outside observer.
   A. correct
  *B. wrong: hidden by the cocoon nebula
   C. wrong: hidden by the master cloud
   D. wrong: hidden by the Bok globule

8. A cocoon nebula surrounds
   A. α-Centauri   D. both A and B
  *B. R Monocerotis   E. both B and C
   C. Sirius

9. _____ are small, highly variable nebulae often found in galactic clusters.
   A. T Tauri stars   D. Hayashi blobs
  *B. Herbig-Haro objects   E. none of these
   C. Bok globules

10. Young, rapidly rotating variable stars surrounded by an expanding gas and dust cloud are
  *A. T Tauri stars   D. Hayashi blobs
   B. Herbig-Haro objects   E. none of these
   C. Bok globules

11. Near young star clusters, it is common to find dark clouds, usually 1000–100,000 AU across with densities at least a million times that of interstellar space. These clouds are called
   A. T Tauri stars   D. Hayashi blobs
   B. Herbig-Haro objects   E. none of these
  *C. Bok globules

12. Objects often found in young star clusters include
   A. T Tauri stars   *D. both A and B
   B. Bok globules   E. all three, A, B, and C
   C. red giants

13. The process of nuclear fusion in which protons directly collide to form deuterium is called the
   A. carbon cycle   D. collision process
  *B. proton-proton cycle   E. direct process
   C. deuterium process

14. Stars of the Sun's mass or less primarily operate on the _____ while they are on the main sequence.
   A. carbon cycle   C. triple-alpha process
  *B. proton-proton cycle   D. none of these

15. The nuclear process that involves collisions between hydrogen and carbon to finally (after a few steps) produce helium is the
    A. direct process
    *B. carbon cycle
    C. proton-proton cycle
    D. triple-alpha process
    E. none of these

16. Stars more massive than 1.5 $M_\odot$ primarily generate their energy while on the main sequence by the
    A. direct process
    *B. carbon cycle
    C. proton-proton cycle
    D. triple-alpha process
    E. none of these

17. The proton-proton cycle dominates when central temperatures are
    A. less than 15 K
    B. less than 15,000 K
    *C. less than 15 million K
    D. greater than 15 million K
    E. greater than 100 million K

18. The carbon cycle dominates when the central temperatures are
    A. less than 15 K
    B. less than 15,000 K
    C. less than 15 million K
    *D. greater than 15 million K
    E. greater than 100 million K

Problem Involving Optional Mathematical Equation from Another Chapter

*19. (Optional Equation VIII) A T Tauri star consists of an A star of $T = 9000$ K and luminosity of 4×10^{28} joules/s and is surrounded by a spherical cocoon nebula at 3000 K. Use the Stefan-Boltzmann law to calculate the size of the nebula. The nebula's radius is _____ AU.
    A. 5.7
    B. 57
    *C. 570
    D. 5700
    E. 390

    (Assume that energy radiated by star = energy radiated by nebula. Use $E = \sigma T^4 A$ with T temperature and A area of nebula, $E = 4 \times 10^{28}$ joules/s [one joule/s = 1 watt].)

Essay

1. Describe three kinds of observational data that are interpreted as indicating present-day star formation. (See pp. 383–384.)
2. Draw an H-R diagram and show the approaching path to the main sequence (simplest model) for a 1 $M_\odot$ and a 10 $M_\odot$, assuming you could see through the cocoon nebula. How would the cocoon nebula modify this? (See Figs. 18-3, 18-4, and 18-5.)

3. List at least three of the four features of young, forming stars that tie in with the evidence we have about the nebula that surrounded our own Sun. (See p. 393.)
4. Describe Herbig-Haro objects. (See p. 393.)
5. Describe T Tauri stars. (See pp. 397–400.)
6. Comment on the old concept of "fixed stars" as permanent, immovable objects on the distant "celestial sphere."

Word Practice

1. The nebula surrounding a forming star is the _____ nebula.

2. Newly formed stars are very bright for short periods, known as the

   _____ _____.

*3. Our solar system is relatively _____, so all stars did not form at the same time.

4. Dense enough interstellar nebula may collapse under the pressure of

   _____ _____.

5. A cloud of gas and dust in space is called a(n) _____.

*6. Fragmentation appears to be an almost _____ consequence of the cloud collapse process.

7. The plural of nebula is _____.

8. _____ globules are small (1000–100,000 AU), dark clouds with a density about a million times that of normal interstellar space.

*9. _____ _____ exist that are younger than the species *Homo-sapiens*.

*10. The high-luminosity phase is often referred to as the _____

   _____.

11. We know that stars are forming _____.

12. _____ _____ _____ are believed to represent a transitional stage between newly formed stars with opaque nebulae and stars that have blown off their cocoons.

13. The theory of _____ _____ seems to have a good basis because all stages of star formation are reasonably well understood.

*14. A(n) _____ _____, such as an open star cluster, shows that star formation occurred recently.

*15. _____ _____ are small, highly variable nebulae often found in clusters.

16. _____ occurs on shrinking protoclusters to form individual stars.

17. _____ dwarfs are faint, small objects radiating their own heat of contraction in the infrared.

*18. The _____ _____ is part of the Orion star-forming region.

19. _____ stars are very faint in the visible range but are bright in the 1–2 micrometer range.

20. The _____ region is a young star-forming region.

*21. _____ is an example of a black dwarf.

22. The _____ _____ nebula is a dark, round cloud that obscures part of the Milky Way in the Southern Hemisphere.

23. In a(n) _____ jet, gas flows out in two narrow streams in opposite directions.

*24. The _____ theory enables us to calculate how collapsing clouds should fragment.

Answers to Word Practice

1. cocoon	13. star formation
2. high-luminosity phase	14. short-lived cluster
3. young	15. Herbig-Haro objects
4. gravitational contraction	16. subfragmentation
5. nebula	17. brown
6. inevitable	18. Orion nebula
7. nebulae	19. infrared
8. Bok	20. Orion
9. short-lived stars	21. Jupiter
10. Hayashi phase	22. Coal Sack
11. today	23. bipolar
12. T Tauri stars	24. Jeans

Chapter 19
Stellar Evolution II: Death and Transfiguration

Answers to Problems in the Text

1. The center contracts and produces more total energy. (Thus, contraction may be viewed as the significant process.) This increased energy flow pushes the outer layers of the star farther out, where they become cooler. The available helium is soon consumed, so the energy flow outward declines and contraction begins again. Teachers should emphasize contraction in the evolutionary sequence.

2. Because they are so bright that we can see them far away, we see an exaggerated sample. The more numerous red main-sequence stars are so faint that we do not see them at such a great distance.

3. Blue-shifted absorption lines around M giants and supergiants and Wolf-Rayet; O and B stars; P Cygni stars; novae; supernovae with visibly expanding nebulae. The expansion of the Crab nebula has been measured photographically.

4. a. Stars of mass less than 1.4 $M_\odot$. Slightly more massive stars may shed enough mass. Thus, the limit might be 6 $M_\odot$ initially.
   b. Medium mass stars 12–30 $M_\odot$ initially.
   c. High-mass stars 30–100 $M_\odot$ initially.
   White dwarf.

5. It burns its fuel much more rapidly and with a process that requires this higher rate. A main sequence B3 has an absolute magnitude of possibly -4 compared with the Sun's +5, so it loses 4000 times as much energy/s but has only 10 times the fuel.

6. When the star collapses, it rotates faster. Even after the outer layers are blown off, the remaining matter keeps shrinking and reaches a higher spin rate.

7. Theorists in the 1930s predicted objects such as neutron stars and black holes long before they could be observed. Without means to test the predictions, these predictions lay forgotten in the journals until observational techniques were available. Accidental discovery of neutron stars led to the realization that the theory could now be tested. This strongly spurred observations in the 1970s. Black holes are not yet completely understood. More observational evidence is needed.

Advanced Problems

8. a. $V_{circ} = \sqrt{GM/R}$; $G = 6.67 \times 10^{-11}$,
   $M = M_\odot = 1.99 \times 10^{30}$ kg
   $R = R_\oplus = 6.38 \times 10^6$ m
   $V_{circ} = 4.56 \times 10^6$ m/s = 4560 km/s
   $\phantom{V_{circ}} = 1.5\%$ speed of light

b. V_{circ} $= \sqrt{2}V_{circ} = 6450$ km/s
$= 2.1\%$ speed of light

c. Both values are $\sqrt{M_\odot/M_\oplus}$ times values for Earth.

$\sqrt{M_\odot/M_\oplus} = 577$

d. White dwarf.

9. $E = \sigma T^4 A$; for star $*$, $L_* = \sigma T_*^4 A_*$.

Evolution to the right on the H-R diagram means T is decreasing.
If $L_* = $ constant and T_* decreases, A_* must increase.

$A_* = $ area, so size is increasing.

10. $L_* = \sigma T_*^4 A_*$

$L_{WD} = \sigma T_{WD}^4 A_{WD} \leftarrow$ white dwarf (WD)

$L_{MS} = \sigma T_{MS}^4 A_{MS} \leftarrow$ main sequence (MS) of same spectral class

$T_{MS} = T_{WD}$, so $L_{WD}/L_{MS} = A_{WD}/A_{MS}$. If $A_{WD} < A_{MS}$, then
$L_{WD} < L_{MS}$, so white dwarf is below main sequence.

11. We are concerned with material that moves at right angles to the line of sight so we can take distance = 500 parsecs = constant. First find how big the cloud must be to resolve it.

$$\frac{\alpha"}{206,265} = d/D \; ; D = 500 \text{ parsecs} = 500 \times 3 \times 10^{16} \text{ m}$$

$$\alpha = \frac{1"}{2}$$

$$d = \alpha D /(2.06 \times 10^5) = \frac{5 \times 3}{2 \times 2.06} \times \frac{10^{18}}{10^5}$$

$$d = 3.47 \times 10^{13} \text{ m}$$

time this takes is d /speed $= \dfrac{3.47 \times 10^{13} \text{ m}}{10^6 \text{ m/s}} = 3.47 \times 10^7$ s

time $= 3.47 \times 10^7$ s $= 1.1$ years
(Remember, 1 year $= \pi \times 10^7$ s.)

12. Use Wien's law, $L = 0.00290/T$ with L in meters and T in Kelvin degrees. For $T = 10^9$ K, we get
$L = 0.0029/10^9 = 2.9 \times 10^{-3} \times 10^{-9} = 2.9 \times 10^{-12}$ m $= 0.0029$ nm
This is X-ray.
Such observation would not be possible on the ground since the atmosphere is opaque. See Chapter 5.

13. For the ideal gas, Kinetic Energy (KE) = $(3/2)kT$, where $k = 1.38 \times 10^{-23}$ in SI units. So KE $= 3.5 \times 10^{-13}$ and $J = (3/2) \times 1.38 \times 10^{-23} T$, which gives
$T = 1.7 \times 10^{10}$ K $= 17$ billion K.
If the faster atoms move at about 4 times average, then since KE $=$
$(1/2)mv^2$, the highest KE is about $16 = 4^2$, so T is about 1 billion K.
We should also consider the possibility that both particles are moving, so the relative velocity would be used. This reduces the temperature by a factor of 2.

147

14. There would be no change in the Earth's orbit or the orbital velocity. At 1 AU the Newtonian universal law of gravity is unchanged, and still $F = GM$ m/r^2. Since the mass is unchanged, so is the force on the Earth and the circular velocity is still

$$\sqrt{\frac{GM}{R}} \, .$$

Sample Test Questions

True-False

1. When a star first moves off the main sequence, it expands. T
2. When a star first moves off the main sequence, it gets cooler. T
3. Stars that become supergiants have a mass greater than 6 $M_\odot$. T
4. A star begins burning helium as soon as it leaves the main sequence. F
5. The helium flash is the turning of the entire star into helium. T
6. FG Sagittae is an example of a giant star. F
7. Most variable stars represent pregiant stages. F
8. The instability strip is a nearly horizontal region of the H-R diagram. F
9. Low-mass (<0.4 $M_\odot$) stars never become variable stars. T
10. The Hertzsprung gap is a vertical band nearly empty of stars. T
11. Nova stars are always single, isolated stars. F
12. White dwarfs are about the size of a planet. T
13. White dwarfs are kept from collapsing by the Pauli exclusion principle. T
14. White dwarfs always have a mass greater than 1.4 $M_\odot$. F
15. A neutron star can be as small as a few kilometers in diameter. T
16. Pulsars may have solid surfaces. T
17. The event horizon surrounds black holes. T
18. Small black holes (<10^{12} kg) may radiate energy. T
19. Cygnus X-1 is believed to contain a black hole. T
20. A star of the Sun's mass may become a black hole. F

Multiple Choice

1. The helium of a star begins to burn when the star is
   *A. at the extreme right top of the giant region
   B. at the extreme left top of the giant region
   C. leaving the main sequence
   D. approaching the main sequence
   E. It never burns.

2. For a star to become a supergiant, its mass must be greater than about _____ $M_\odot$.

   A. 0.1
   B. 0.6
   C. 1
   *D. 6
   E. 16

148

3. Most stars become variable stars _____ the redgiant stage.
   A. before
   *B. after
   C. instead of going through
   D. during
   E. between

4. The star(s) _____ is (are) an example of a variable star.
   A. Mira
   B. Polaris
   C. Sirius
   *D. both A and B
   E. all three, A, B, and C

5. Which is the correct order of evolutionary stages for a medium-mass star?
   A. main sequence, black dwarf
   B. main sequence, variable star, giant, white dwarf
   *C. main sequence, giant, variable star, white dwarf
   D. main sequence, giant, white dwarf, variable star
   E. none of these

6. The line on the diagram that indicates the instability strip is __B__.

7. If one of the lines on the diagram is the instability strip, the Hertzsprung gap is __C__.

8. The variability of Cepheids is caused by
   A. ionized hydrogen
   B. doubly ionized hydrogen
   C. ionized helium
   *D. doubly ionized helium
   E. none of these

9. Hot stars resembling O stars except for their expanding gaseous shells (shown by ionized carbon, nitrogen, and helium) are called
   A. M-giants
   B. supergiants
   *C. Wolf-Rayet stars
   D. main sequence
   E. novae

10. An example of a fast-evolving supergiant is
    A. M5
    B. the Crab nebula
    C. Mira
    *D. FG Sagittae
    E. Cygnus X-1

11. A nova is caused by a member of a binary pair becoming a giant and transferring matter to its more evolved companion. The more evolved companion then becomes a nova.
  *A. correct
  B. wrong: Transfers are to less-evolved star.
  C. wrong: Less-evolved star is one that undergoes nova.
  D. wrong: Both B and C are needed.
  E. misleading: Nova stars are single stars.

12. The Chandrasekhar limit is
  A. the upper limit of size for all stars
  B. the upper limit of mass for all stars
  *C. the upper limit of mass for white dwarfs
  D. the lower limit of mass for white dwarfs
  E. the maximum mass a star can have without becoming variable sometime in its life

13. The Chandrasekhar limit is _____ $M_\odot$.
  A. 0.8   D. 6
  *B. 1.4   E. 59
  C. 2.3

14. Supernovae can be as bright as magnitude
  A. 20   D. -3
  B. 3   *E. -20
  C. 0

15. Supernovae explosions are caused by
  *A. neutrino absorption   D. gas loss
  B. proton absorption   E. star collisions
  C. electron absorption

16. The period of the Crab nebula pulsar is about
  A. 0.33 years   D. 3.3 seconds
  B. 0.33 months   E. 33 seconds
  *C. 33 milliseconds

17. While a supernova may increase a star's luminosity a few hundred million times, a nova is even brighter.
  A. correct
  B. wrong: A nova is about as bright as a supernova.
  *C. wrong: A nova is much less bright than a supernova.
  D. wrong: A supernova increases a star's luminosity only three or four times as much.
  E. wrong: Both C and D are correct.

18. The supernova remnant for stars with masses between 12 and 30 $M_\odot$ is most likely
    A. a G2 star
    B. a white dwarf
    C. a black hole
    *D. a neutron star
    E. an exohole

19. A good analogy of how a neutron star is a pulsar is
    A. a blinking neon sign
    *B. a lighthouse
    C. a greenhouse
    D. a metronome
    E. a traffic light

20. A _____ is so dense that its gravitational field keeps light from escaping.
    A. white dwarf
    B. supergiant
    C. neutron star
    *D. black hole
    E. shell star

21. The surface surrounding a black hole where the escape velocity is the speed of light is the
    A. photon sphere
    B. singularity
    *C. event horizon
    D. last surface
    E. Einstein-Rosen bridge

22. The periods of variation in light output of certain classes of _____ are known to be related to their absolute magnitudes.
    A. binary stars
    *B. variable stars
    C. white dwarfs
    D. red stars
    E. none of these

luminosity

A.
period (days) 1   100

luminosity

B.
period (days) 1   100

luminosity

C.
period (days) 1   100

luminosity

D.
period (days) 1   100

23. Refer to the diagram above. Of the plots in the diagram, __B__ shows the period-luminosity behavior for Cepheids.

Essay

1. Describe why and how Cepheid variables oscillate. (See pp. 409–410.)
2. List and describe three types of stars that blow off mass. (See pp. 412–413.)
3. Describe why neutron stars pulse. (See pp. 424–425.)

4. Describe some of the paradoxical properties of black holes. (See pp. 425–428.)
5. How were the early analysis of white dwarfs and the prediction of neutron stars historically related to the atom bomb? (See p. 430.)
   (Main point: The studies of high-density jamming together of atoms as a part of nuclear physics led to developments in both astrophysics and weapons technology.)
6. Discuss how evolved stars may lose mass. (See pp. 411–418.)

Word Practice

1. A(n) _____ _____ is a planet-sized, superdense star.

*2. The _____ _____ is a vertical band on the H-R diagram that is nearly empty of stars.

3. The _____ _____ _____ keeps the electrons in a white dwarf from collapsing further.

4. The stars of 1–5 $M_\odot$ form _____ _____ after they leave the main sequence.

5. The _____ _____ gives a value for the upper mass limit of white dwarfs.

6. The _____ effect describes how stars in the main sequence move toward the giant region.

7. A(n) _____ is a pulsating radio source believed to be a rapidly rotating neutron star.

*8. The variations in Cepheid stars are caused by a(n) _____ layer of doubly ionized helium.

9. _____ holes are surrounded by an event horizon.

*10. _____ _____ often have blue-shifted absorption lines, showing that they have blown off dusty gas.

*11. _____ _____ and gamma rays may be produced by material falling into black holes.

12. _____ _____ have regular variations in light with periods related to the intrinsic brightness.

*13. Doubly _____ _____ provides a dam that drives the Cepheid variables.

14. _____ stars are similar to O stars except for their expanding gaseous shells.

15. A star that varies irregularly is called a(n) _____ _____.

152

16. When matter is squeezed so densely that only the electrons provide the support, we have _____ _____.

17. When the gravitational pressure in a star is able to overcome the electric repulsion and collapse the atoms, a(n) _____ _____ is formed.

18. A star with more than 6 $M_\odot$ will evolve into a(n) _____ as it leaves the main sequence.

19. The _____ _____ is a near-vertical region of the H-R diagram in which Cepheid and RR Lyrae variables are found.

20. A rapidly rotating neutron star is visible as a(n) _____.

21. A(n) _____ _____ is a star that varies in brightness.

*22. _____ _____ is the fastest-evolving supergiant known.

23. A(n) _____ _____ results when a star becomes so small that even light cannot escape.

24. The _____ _____ around a black hole is defined as that surface from which no radiation can escape.

25. The _____ is the most energetic of all stellar explosions.

Answers to Word Practice

1. white dwarf
2. Hertzsprung gap
3. Pauli exclusion principle
4. giant stars
5. Chandrasekhar limit
6. funneling
7. pulsar
8. opaque
9. black
10. M giants
11. X-ray sources
12. Cepheid variables
13. ionized helium
14. Wolf-Rayet
15. irregular variable
16. degenerate matter
17. neutron star
18. supergiant
19. instability strip
20. pulsar
21. variable star
22. FG Sagittae
23. black hole
24. event horizon
25. supernova

Chapter 20
Interstellar Atoms, Dust, and Nebulae

Answers to Problems in the Text

1. The thicker atmosphere and dust are able to scatter the blue light better than the red. The blue has been bent away from the line of sight. Looking away from the Sun, we see that scattered light.

2. Because the dust on Mars is red, it reflects red light better than blue.

3. Two ways: (1) Strong radiation pressures and stellar winds stir up nearby interstellar material. (2) They lose mass and even explode as supernovae.

4. More and more carbon compounds of greater and greater complexity are being found—even some that could make biologically significant compounds. So carbon chemistry seems the same elsewhere in the universe.

5. (See Table 20-1.) Masses range from 0.1 $M_\odot$ to 100,000 $M_\odot$. Since stars form in clusters, we expect those nebulae associated with star formation to have masses hundreds to thousands of times that of the Sun. Dark and emission nebulae behave this way.

6. Emission nebulae must, by definition, be near a hot star to be excited enough to have emission lines. Solar-type stars are not hot enough to cause emission nebulae over a significant volume of space.

7. Planetary nebulae are expanding. They are bubblelike shells thrown off by stars, perhaps in a cyclic sequence, whereas emission nebulae are chaotic gas clouds within which hot, young, violent stars are forming.

Advanced Problems

8. N = total number of atoms = $(4/3)\pi r^3 n$, where n = number per unit volume and r = size. Take r = 10 atoms/cm^3, $N = 10^{57}$ atoms; then

$$r^3 = \frac{3N}{4\pi n} = \frac{3 \times 10^{57}}{4 \times 3.14 \times 10/cm^3} = 2.38 \times 10^{55} \text{ cm}^3$$

$$r = 2.88 \times 10^{18} \text{ cm} = 0.96 \text{ parsecs}$$

a. Much larger than the solar system, by a factor of 10,000 in linear size.

b. Small compared to the dark and emission nebulae in Table 20-1, all of which (except 1) have a larger density.

9. a. $0.005 = v/c = v/(3 \times 10^8 \text{ m/s})$

$$v = 1.5 \times 10^6 \text{ m/s} = 1500 \text{ km/s}$$

b. $D = 1.5 \times 10^3$ parsecs = 4.5×10^{19} m in one year $\alpha = 0''2$

$$\alpha/2.06 \times 10^5 = d/D$$

$$d = (\alpha D)/2.06 \times 10^5 = 9 \times 10^{18}/2.06 \times 10^5$$

$$= 4.36 \times 10^{13} \text{ m}$$

So, in one year, it expands 4.36×10^{13} m; hence, in one second

$$v = \frac{4.36 \times 10^{13} \text{ m}}{\pi \times 10^7 \text{ s}} = 1.39 \times 10^6 \text{m/s}$$

$$= 1390 \text{ km/s}$$

    c. Agree to within 11 parts in 150 = 0.073. Agree to within 7.3%.

10. Use escape velocity = $(2GM/R)^{1/2}$, where $M = 100$ solar masses and R is the distance from the center. We also have

      $(1/2)\ mv^2 = (3/2)kT$, where $m = 1.67 \times 10^{-27}$ kg.

Then

      $v = (3kT/m)^{1/2} = 1.58 \times 10^5$ m/s,

while the escape velocity = 210 m/s. Thus, the cloud clearly continues to expand.

Sample Test Questions

True-False

1. The interaction of any wavelength of radiation with matter depends only upon the wavelength and is independent of the size of the particles making up the matter. F
2. An atom is excited if one or more electrons are knocked completely out of the atom. F
3. When molecules absorb light, they affect a range of wavelengths (bands) rather than discrete, individual wavelengths. T
4. The sky of the Earth is blue because more blue than red light is scattered by air molecules. T
5. The most common gas in interstellar space is sodium. F
6. Metastable states are states of atoms with unusual electron distribution that last for several minutes. T
7. Metastable states give rise to forbidden lines. T
8. The gases in space are too rare for molecules to form. F
9. The 21-cm radiation can be emitted by atomic hydrogen. T
10. Interstellar obscuration is caused by hydrogen gas. F
11. The interstellar grains (dust) are from 0.0001 to 0.011 mm in size. T
12. The interstellar dust is spherical in shape. F
13. The redder the light, the more energy each photon carries. F
14. In HII regions, the hydrogen gas is neutral. F
15. The Orion nebula is an emission nebula. T
16. At 1000–2000 nm, the brightest object in the sky is Eta Garinae. T
17. The Gum nebula has an angular diameter of 6°. F

Multiple Choice

1. The interaction between radiation and matter depends upon
   A. the wavelength of radiation
   B. types of particles
   C. sizes of particles
   D. both A and C
   *E. all three, A, B, and C

The following diagram is for questions 2–4.

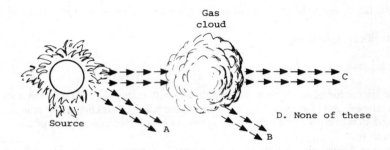

2. Where in the diagram will you see absorption lines added to the star's spectrum by the gas cloud? _C_.

3. Where in the diagram will you see bright emission lines from the gas cloud? _B_.

4. Where in the diagram will you see the source reddened by the gas cloud? _D_.

   (*Note:* Gas clouds do not cause reddening.)

The following diagram is for questions 5–6.

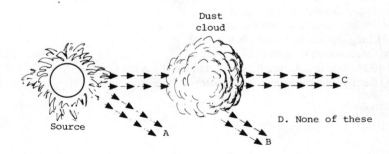

5. Where in the diagram will you see only redder light from the star? _C_.

6. Where in the diagram will you see the cloud glowing bluish? _B_.

156

7. Calcium and _____ provide the major absorption lines in HI regions.
   A. helium
   B. oxygen
   C. gold
   *D. sodium
   E. uranium

8. The 21-cm radiation from HI regions happens when
   A. two molecules collide
   B. two atoms collide
   *C. the spin of the electron in a hydrogen atom flips with respect to the spin of the nucleus
   D. an O or B star is near
   E. both B and D occur

9. In addition to interstellar gas and dust, five interstellar molecules have been found.
   A. correct
   B. wrong: no gas known
   C. wrong: no dust known
   *D. wrong: over 30 known
   E. wrong: both C and D

10. One of the interstellar molecules found is an amino acid.
    A. correct
    B. wrong: Two are.
    C. wrong: Three are.
    D. wrong: But the spectrum of bacteria has been found.
    *E. wrong: No amino acids have been found, only two molecules that could make an amino acid.

11. The interstellar grains (or dust) are
    *A. small, smokelike particles
    B. all the same size
    C. spherical
    D. both A and C
    E. all three, A, B, and C

12. Planck's law relates
    A. the change in wavelength to radial velocity
    B. the break in radiation to temperature
    *C. the wavelength of light to energy per photon
    D. the luminosity to temperature and source size
    E. none of these

13. The bluer the light, the _____ each photon contains.
    *A. more energy
    B. less energy
    C. more wavelength
    D. more speed
    E. none of these

14. HI regions contain mainly ionized hydrogen, while HII regions contain mainly neutral hydrogen.
    A. correct
    B. wrong: Replace hydrogen with helium.
   *C. wrong: Interchange HI and HII.
    D. wrong: Both B and C changes are needed.
    E. wrong: HI regions contain mainly helium dust, and HII regions contain mainly helium gas.

15. To exist, HII regions need
    A. an M star
    B. a source of radiation longer than 91.2 nm
   *C. a source of radiation shorter than 91.2 nm
    D. sodium atoms
    E. both A and B

16. Clouds that look brighter than the background by virtue of emitted or scattered light are called _____ nebulae.
    A. planetary   D. Crab
    B. dark   E. emission
   *C. bright

17. Clouds rich in dust silhouetted against a bright background region are called _____ nebulae.
    A. planetary   D. reflection
   *B. dark   E. emission
    C. bright

18. Several bright stars seem to have been ejected at high speed from the Orion nebula. They are called runaway stars.
   *A. correct
    B. wrong: called ejected stars
    C. wrong: seem to be approaching the nebula, not leaving
    D. wrong: called injected stars
    E. wrong: both C and D are needed

19. Which nebula has the largest angular diameter as seen from the Earth?
   *A. Gum   D. Pleiades
    B. Orion   E. Crab
    C. Eta Carinae

20. (Optional Equation I) The Crab nebula was seen about 925 years ago (1 year $\approx \pi \times 10^7$ s). It is about 2200 parsecs from us and appears at 4.7 minutes of arc. The linear size of the nebula (when the light we now see left the nebula) was _____ parsecs.

A. 0.1   D. 30
B. 0.3   E. 100
*C. 3

21. (Optional Equation VII) In the last 925 years (approximately) since the Crab nebula was first seen, the nebula has appeared to expand at an average rate of 1550 km/s. Neglecting the turbulence of the expansion, what should the average Doppler shift be in the light from the outer parts of the nebula in the side toward us?

*A. 0.5% blue   D. 5% red
B. 0.5% red   E. no shift
C. 5% blue

Essay

1. Describe the two main principles that govern what happens when light from a star passes through clouds of interstellar atoms, molecules, and dust grains. (See p. 436.)
2. Explain why the sunset is red and the sky is blue. (See pp. 439–440.)
3. List the five main properties of the interstellar grains. (See p. 444.)
4. Name and describe three of the four types of interstellar regions. (See p. 444.)
5. Describe some of the interesting properties of the Orion nebula. (See pp. 449–454.)
6. Describe some of the interesting properties of the Eta Carinae nebula. (See pp. 454–457.)

Word Practice

1. _____ _____ are bright nebulae close enough to a hot star that ionization or excitation has occurred.

2. The _____ _____ is an effect of dust that makes objects look redder.

3. _____ nebulae are bright nebulae that are near enough to a star to reflect measurable light.

4. _____ scattering causes the sky to appear blue.

5. Entries in the New General Catalog are identified by _____

_____.

6. _____ of an interstellar gas occurs when electrons are knocked to a higher energy level.

7. _____ _____ are interstellar clouds that look brighter than the background.

8. _____ nebulae are tight, expanding spherical shells of gaseous material.

9. The _____ _____ appears to have an angular diameter of 60°. At its center lies the Vela pulsar.

10. High-speed stars racing out from the center of the Orion nebula are called _____ _____.

11. _____ nebulae are clouds rich in dust silhouetted against bright background clouds or against a star-rich region.

12. Regions where hydrogen is neutral are called _____ regions.

13. A cloud in space is called a(n) _____.

14. Cold regions (~ 10 K) of dark clouds of dust and gas are _____ _____.

15. _____ _____ such as sodium produce prominent absorption lines.

16. _____ occurs when electrons are knocked out of atoms.

17. _____ _____ causes the red appearance of the Sun at sunset.

18. Interstellar _____ cause reddening and general dimming of starlight.

19. _____ _____ tells us that hotter sources emit more blue light than cooler sources.

20. Interstellar grains cause a general dimming of starlight called _____ _____.

21. The _____ _____ _____ is a strange emission nebula near the Coal Sack region.

22. _____ _____ arise from unusual states of interstellar atoms.

23. _____ _____ were assigned in 1781 to nonstellar objects, including the particularly vivid nebulae.

24. The _____ _____ is a star-forming region about 460 pc from us.

25. _____ states are possible for interstellar atoms because the density of material is so low.

26. _____ regions contain ionized hydrogen.

27. Neutral hydrogen emits radiation at about _____ wavelength.

28. _____ _____ identify nebulae in the Index Catalog.

29. _____ law relates photon energy to light wavelength.

Answers to Word Practice

1. emission nebulae
2. interstellar reddening
3. reflection
4. Rayleigh
5. NGC numbers
6. excitation
7. bright nebulae
8. planetary
9. Gum nebulae
10. runaway stars
11. dark
12. HI
13. nebula
14. molecular clouds
15. interstellar atoms
16. ionization
17. Rayleigh scattering
18. grains
19. Wien's law
20. interstellar obscuration
21. Eta Carinae nebula
22. forbidden lines
23. Messier numbers
24. Orion nebula
25. metastable
26. HII
27. 21-cm
28. IC numbers
29. Planck's

Chapter 21
Companions to Stars:
Binary, Multiple, and Possible Planetary Systems

Answers to Problems in the Text

1. Space program, binary stars. In 1827, Xi Ursa Majoris, a binary in Ursa Majoris, was found to fit a Keplerian ellipse.
2. Often highly elliptical orbits in binaries. In multiples, the relative inclination of orbits may vary; that is, for most multiple star systems (open), there is no common *plane of the system* . The masses of companions in known star systems are "of course" much greater than planetary masses.
3. Close binaries have a different mass distribution than open binaries. The text mentions one coplanar triple star system, possibly related to a planetary system.
4. All known novae have occurred in either known or suspected binary systems. Apparently, mass transfer moves the smaller star into the nova state. Supernovae are not related to binaries.
5. The 3 $M_\odot$ will evolve faster and transfer some mass to the 1 $M_\odot$ star. This may make the 1 $M_\odot$ star unstable. The 1 $M_\odot$ star, now having extra mass, will evolve faster than before. When the 1 $M_\odot$ star (now possibly more massive) becomes a giant, matter is transferred to the 3 $M_\odot$, which may become a nova.
6. The matter density would be higher in a region of star-cluster birth. Thus, that nearby stars would form—or protostars fission—is much likelier in clusters.

Advanced Problems

7. Very nearly that of the Earth. One approach is to use Kepler's third law as modified by Newton:
    $$(M_1 + M_2)P^2 = r^3,$$
    where M = mass in $M_\odot$, P = period in years, and r = distance in AU.
    For Earth, $(1 + \text{small})1^2 = 1^3$.
    For $M_2 = 0.05$, $(1 + 0.05)P^2 = 1^3 \rightarrow P^2 = \dfrac{1}{1.05}$
    $P = 0.985$ years
    Students may either neglect the 0.05 and get $P = 1$ year or use $V_{circ} = \sqrt{GM/R}$ and period $= 2\pi/V$ and again get $P = 1$ year.
8. $V_{circ} = \sqrt{GM/R}$; $G = 6.67 \times 10^{-11}$, $M = 2 \times 10^{30}$ kg
    $R = 7.8 \times 10^{11}$ m
    $V_{circ} = 1.31 \times 10^4$ m/s = 13.1 km/s
    a. This agrees with the velocity for Jupiter (as it must).
    b. $\sqrt{2} \times 13.1$ km/s = 18.5 km/s
9. Since $5M \gg 0.5M$,
    $V_{circ} = \sqrt{GM/R}$; $G = 6.67 \times 10^{-11}$, $M = 5 M_\odot = 10^{31}$kg, so $R = GM/V^2$
    $(4.7 \times 10^4 \text{ m/s})^2 = GM/R$

$$R = \frac{6.67 \times 10^{-8} \times 10^{31}}{(4.7)^2 \times 10^8} \text{ m} = 3.02 \times 10^{11} \text{ m} = 2 \text{ AU}.$$

If we use Kepler's law as we did in question 7,

$$(5 + .05)P^2 = R^3$$

$$P = 2R/V \quad (V \text{ in AU/year})$$

$$5.5 \frac{4\pi^2 R^2}{V^2} = R; \quad R = \frac{5.5 \times 4\pi^2}{V^2}$$

$$V = 4.7 \times 10^4 \text{ m/s} \times \frac{\text{AU}}{1.5 \times 10^{11} \text{ m}} \times \frac{\pi \times 10^7 \text{ s}}{\text{year}}$$

$$V = 9.85 \text{ AU/year}$$

$$R = \frac{5.5 \times 4 \times \pi^2}{(9.85)^2} = 2.25 \text{ AU}$$

Expect the first method to have an error of $(0.5)/5 \sim 10\%$, which is about the difference given by the two approaches.

10. Since the problem does not give a radius, we can give only relative data.
    a. $v = (GM/R)^{1/2}$
    Let the orbital radius be R AU and $M = 10$ solar masses, so $v = (10/R)^{1/2} \times V$, where V is the Earth's orbital velocity.
    b. $F = GMm/R^2$. Since M changes from 10 to 4 solar masses, $F_{\text{afterward}}/F_{\text{before}} = .4$.
    c. Neglect any direct forces from the explosion. Then from part a, the new orbital speed for circular orbit is $(4/10)^{1/2} = .64$ of the old, so it is going too fast for circular motion and hence it probably escapes.

Sample Test Questions

True-False

1. All optical doubles are true binaries. F
2. The first discovery of gravitational orbital motion beyond the solar system was Xi Ursa Majoris. T
3. The brighter star in a binary is denoted by the letter B for brighter. F
4. If orbital motion of a pair of stars is revealed by periodic Doppler shifts, we have a Doppler binary. F
5. A plot of brightness versus time is called a light curve. T
6. Binary systems revealed by the superposition of spectral lines from stars of two different temperatures are astrometric binaries. F
7. All estimates and counts give more than 50% of all stars as single stars. F
8. Close binaries are those binary systems that are close to us. F
9. Close binaries may be related to planetary systems. T
10. The material in each lobe of the Lagrangian surface is bound to the star in the other lobe. F
11. In contact binaries, both lobes of the Lagrangian surface are filled. T
12. One-half of all novae are believed to be binary systems. F
13. W Ursa Majoris stars are contact binaries. T

163

14. The only way binary or multiple-star systems can form is by common condensation. F

Multiple Choice

1. A gravitationally associated pair of stars, both of which can be separately observed, is called a(n) _____ binary.
   A. optical
   B. spectrum
   C. eclipsing
   D. astrometric
   *E. visual

2. If what appears to be a single star undergoes wavy motions, we know it as a(n) _____ binary.
   A. visual
   B. spectroscopic
   C. optical
   *D. astrometric
   E. spectrum

3. Spectroscopic binaries show Doppler shifts. This means we measure their
   A. different temperatures
   B. colors
   *C. radial velocities
   D. tangential velocities
   E. separation

4. If the spectrum of a star shows spectral lines both of a hot and of a cool star, we have a(n) _____ binary.
   A. spectroscopic
   *B. spectrum
   C. visual
   D. eclipsing
   E. none of these

5. A reasonable estimate of the percentage of stars that are multiple is _____ %.
   A. 3
   B. 5
   C. 10
   D. 20
   *E. 60

6. Most multiple stars differ from planetary orbits in that
   A. orbits are often highly elliptical
   B. orbits have all values of inclination
   C. they obey Kepler's laws
   *D. both A and B
   E. all three, A, B, and C

7. Which of the following figures best represents the Lagrangian surface for a binary system of equal mass?   _A_ .

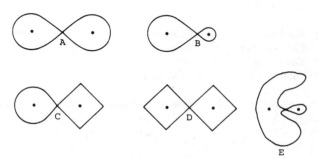

8. Which of the above figures best represents the Lagrangian surface for a binary system with the stars of unequal mass?   _B_ .

9. In a contact binary, _____ Lagrangian lobe.
   A. one star fills its
   B. neither star fills its
  *C. both stars fill their
   D. only one star has a
   E. none of these

10. There is evidence that all supernovae are members of binary systems.
   A. correct
   B. wrong: only about 50%
  *C. wrong: novae, not supernovae
   D. wrong: both B and C changes needed

11. Only the evolution of the more massive star in a close binary system is affected by the presence of the other star.
   A. correct
   B. wrong: Only the less massive is affected.
   C. wrong: The only effect is in triple systems.
  *D. wrong: Both are evolutionarily affected.
   E. wrong: Neither is evolutionarily affected.

12. Which of the following is a contact binary?
  *A. W Ursa Majoris   D. Barnard's star
   B. Cygnus X-1   E. none of these
   C. Alpha Centauri

13. In Algol, the originally more massive star has transferred matter to the originally less massive. Thus, the more massive star is now on the main sequence and the less massive is a white dwarf.
    A.   correct
    B.   wrong: The less massive has transferred matter to the more massive.
    C.   wrong: The more massive is a white dwarf.
    *D. wrong: The less massive is a giant star.
    E.   wrong: Both C and D changes are needed.

14. Which of the following is a binary system that is also a strong X-ray source?
    A.   Ursa Majoris X-3   D.   W Ursa Majoris
    *B. SS-433   E.   none of these
    C.   Algol

15. Formation of binary stars by chance encounter is the principle behind the
    _____ theories of binary star formation.
    *A. capture   D.   encounter
    B.   fission   E.   van de Kamp
    C.   condensation

Essay

   1. Give four of the six types of physical binary star systems classified according to the method of detecting them. For each, give the detection method. Discuss whether these types are physically different in origin, and so on. (See pp. 463–464.)
   2. Discuss the estimates of the incidence of multiplicity. (See p. 468.)
   3. Give the main stages in the evolutionary path of a close binary that at first had neither star filling its Lagrangian lobe but ended up with both lobes filled. Assume the masses were initially very unequal. (See pp. 468–469.)
   4. Briefly describe and name the three groups of theories of binary star origin. (See pp. 478–479.)

Word Practice

*1. _____ are the results of binary star evolution.

2. The _____ _____ _____ is a measure of the probability of a star's belonging to a multiple system.

3. If observational effects tend to make observational data nonrepresentative of the whole population, we have _____ _____.

4. In _____ binary systems, our line of sight is such that the stars pass in front of each other.

5. In _____ binary systems, the two stars are touching and share a common envelope.

6. The _____ theory proposes that binaries form by one star splitting into two.

7. A plot of luminosity versus time for a binary system is called a(n) _____ curve.

8. A pair of co-orbiting stars is called a(n) _____ star.

9. _____ _____ _____ stars are contact binaries.

10. In _____ _____, both stars fill their Lagrangian lobes.

11. _____ double stars are not co-orbiting.

12. Eclipses in binary systems are detected by plotting the _____ _____.

13. _____ _____ are binaries in which both stars fill their Lagrangian lobes.

14. _____ binaries are revealed by motions measured with respect to background stars.

15. _____ _____ binaries show both Doppler shift and eclipses.

16. A(n) _____ _____ of binary stars would have the two stars forming separately and later becoming gravitationally bound.

17. In _____ binaries, we can resolve both stars.

18. As binary stars age, matter can be transferred across the Lagrangian surfaces, leading to different _____ than for single stars.

19. _____ binaries are detected by observing spectral lines characteristic of stars at two different temperatures.

20. In _____ _____ binaries, the stars pass in front of each other and have measurable Doppler shifts.

21. The _____ surface separates the regions where matter is bound to the individual stars of a binary from the region where the matter is in common.

22. A(n) _____ _____ consists of two stars gravitationally bound.

23. The Lagrangian surface of a binary system has two _____.

24. _____ star systems contain three or more stars.

*25. Xi Ursa Majoris was the first binary system shown to obey _____'s laws.

*26. _____ is an unusual eclipsing binary in that the less massive companion is the more evolved giant.

27. In _____ binaries, orbital motion is detected by periodic Doppler shifts in spectral lines.

28. According to _____ theories of binary formation, the stars of a binary system formed from separate subfragments of a single cloud.

Answers to Word Practice

1. novae
2. incidence of multiplicity
3. selection effects
4. eclipsing
5. contact
6. fission
7. light
8. binary
9. W Ursa Majoris
10. contact binaries
11. optical
12. light curve
13. contact binaries
14. astrometric
15. eclipsing spectroscopic
16. capture theory
17. visual
18. evolution
19. spectrum
20. eclipsing spectroscopic
21. Lagrangian
22. physical binary
23. lobes
24. multiple
25. Kepler
26. Algol
27. spectroscopic
28. subfragmentation

Chapter 22
Star Clusters and Associations

Answers to Problems in the Text

1. O and B stars are young stars. Open clusters are young enough that the O and B stars have not "burned out." Since they are massive, they are hot and so burn their hydrogen rapidly, at a great rate, making them the brightest main-sequence stars. Globular clusters are old enough that the O and B stars have run out of fuel and become faint stars, such as supernova remnants. The next most massive stars have evolved to giants.

2. Globular clusters (so many stars).
   Globular clusters.
   Globular clusters.

3.

Open Clusters   Globular Clusters

Associations

4. We might see 10,000 stars as close as, if not closer than, the Centaurian system. The sky would be a lively spectacular at night. If we were at 1 AU from a G star, it would still be easy to tell night from day. However, many globular clusters are X-ray sources, so internal conditions might be bad for us.

Advanced Problems

5. Volume with diameter 4 parsecs $\rightarrow V = \frac{4}{3}\pi r^3 = \frac{4}{3}\pi \left(\frac{4}{2}\right)^3 = 33.5$ parsecs^3. In 33.5 parsecs^3, there are 350 stars $\rightarrow$ 10.4 stars/parsecs^3, so there is a volume of 1/10.4 parsecs^3 per star. Set this equal to a volume and find the size of the volume. Make the volume a cube; a side of the cube is separation of

$$\left(\frac{1}{10.4}\right)^{1/3} = (9.62 \times 10^{-2})^{1/3} \approx 0.5 \text{ parsecs.}$$

In solar neighborhood, about 1 parsec separation. About a factor of 2 in separation or $2^3 = 8$ in density.

6. Again, $V = \frac{\pi d^3}{6} = 114$ parsecs^3 or $\frac{2 \times 10^5}{114} = 176$ stars/parsec^3.

Separation $= \left(\frac{1}{176}\right)^{1/3}$ parsec $= 0.178$ parsec.

Near the Sun, 1 parsec separation.
About a factor of 5.6 in separation or $(5.6)^3 = 176$ in density.

7. $\alpha''/206{,}265 = d/D$; $D = 10^3$ parsecs; $\alpha'' = 1''$.

   a. $d = (\alpha''D)/(2.06 \times 10^5) = 4.9 \times 10^{-3}$ parsecs ~ 1000 AU

   b. $D = 10 \times D$ of part a, so $d = 4.9 \times 10^{-2}$ parsecs $= 10{,}000$ AU

8. a. $V_{circ} = \sqrt{GM/R}$; $G = 6.67 \times 10^{-11}$, $M = 6 \times 10^{35}$ kg

     $R = 1.5 \times 10^{17}$ m

$$
\begin{aligned}
V_{circ} &= [(6.67 \times 6/1.5) \times (10^{-11} \times 10^{35}/10^{17})]^{1/2} \\
&= (26.7 \times 10^7)^{1/2} \\
&= 1.6 \times 10^4 \text{ m/s} \\
&= 16 \text{ km/s}
\end{aligned}
$$

b. Look for a Doppler shift in light from the stars at the edges of the cluster. But if the axis of revolution around the cluster points toward the Earth, it may be hopeless! If the orbital plane is properly oriented, shift is

$$\frac{V}{C} = \frac{16}{3 \times 10^5} = 5.4 \times 10^5,$$

so the change in wavelength is about 0.005% of the original wavelength, which might be possible to detect.

9. The small angle equation, $\dfrac{\alpha''}{2.1 \times 10^5} = \dfrac{d}{D}$ requires α'' to be in seconds of arc. The cluster measures about $1°4 = 4.8 \times 10^3{''}$. Then

$$d = \frac{4.8 \times 10^3}{2.1 \times 10^5} \times D = \frac{4.8 \times 10^3}{2.1 \times 10^5} \times 127 \text{ parsecs} = 3.0 \text{ parsecs,}$$

which compares favorably with the figure of 2 parsecs quoted in Fig. 22-3.

Sample Test Questions

True-False

1. Moderately close-knit, irregularly shaped groupings of stars are called globular clusters. F
2. Associations are called O associations or B associations, depending on whether the brightest stars are O or B. F
3. The Pleiades are an example of an open cluster. T
4. Cluster distances are determined by measuring parallaxes. F
5. Cepheid variables can be used as distance indicators. T
6. The open clusters lie in a disk in our galaxy. T
7. The globular clusters form a spherical cloud in our galaxy. T
8. About half of the identified open clusters are less than 100 million years old. T
9. Most open clusters are older than globular clusters. F
10. Most associations are more than 1 billion years old. F
11. Globular clusters are more than 12 billion years old. T
12. A sphere is the final stable state for a contracting, rotating, self-gravitating system with substantial initial rotation. F
13. Globular clusters should last less than 20 million years before breaking up. F

Multiple Choice

1. Moderately close-knit, irregularly shaped groupings of stars 4 to 20 parsecs in diameter are called
   A. groupings
   *B. open clusters
   C. globular clusters
   D. associations
   E. clumps

2. Loose-structured, irregularly shaped groupings of stars very rich in O and B or T Tauri stars are called
   A. groupings
   B. open clusters
   C. globular clusters
   *D. associations
   E. clumps

3. Massive, tightly packed spherical, symmetrical groupings of stars are called
   A. groupings
   B. open clusters
   *C. globular clusters
   D. associations
   E. clumps

4. Some (or one) of the methods of measuring cluster distances are (is) by
   A. parallaxes
   B. star luminosities
   C. cluster diameters
   *D. both B and C
   E. all three, A, B, and C

5. Interstellar dust throws off estimates of cluster distances since it dims the stars. This effect can be corrected by measuring the amount of
   A. Doppler shift
   B. interstellar bluing
   *C. interstellar reddening
   D. both A and B
   E. both A and C

6. The most useful method of determining the distances of distant clusters is to use
   A. parallaxes
   *B. Cepheid variables
   C. star luminosities
   D. cluster diameters
   E. Doppler shifts

7. Which clusters lie in a spherical distribution about the center of our galaxy?
   A. associations
   *B. globular
   C. open
   D. peanut
   E. both A and C

8. Which clusters lie nearly in a disk (the plane of the galaxy)?
   A. associations
   B. globular
   C. open
   D. peanut
   *E. both A and C

171

9. A reasonable range for the ages of open clusters is _____ years.
   A. 1–30 million   D. 1–10,000
   *B. 1–100 million   E. 10–40 billion
   C. 12–13 billion

10. A reasonable range for the ages of associations is _____ years.
    *A. 1–30 million   D. 1–10,000
    B. 1–100 million   E. 10–40 billion
    C. 12–13 billion

11. A reasonable range for the ages of globular clusters is _____ years.
    A. 1–30 million   D. 1–10,000
    B. 1–100 million   E. 10–40 billion
    *C. 12–13 billion

The following diagrams are for questions 12–14.

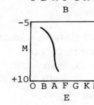

12. Which diagram above best represents the H-R diagram for an association? _B_.

13. Which diagram above best represents the H-R diagram for an open cluster? _A_.

14. Which diagram above best represents the H-R diagram for a globular cluster? _D_.

15. Globular clusters are spherical rather than flattened because the initial angular momentum was
    A. high   D. clockwise
    B. moderate   E. counterclockwise
    *C. low

16. Globular clusters are sometimes X-ray sources. One (or more) possible explanations is (are)
    A. many neutron stars and black holes colliding
    B. high-mass black holes disrupting stars
    *C. neutron stars and black holes capturing binary members and then transferring mass
    D. both A and B
    E. all three, A, B, and C

Essay

1. Name and describe the three types of clusters. (See pp. 481–484.)
2. What are the three methods of determining cluster distances? Which is the best? (See pp. 485–487. *Note:* Parallax is not a method used for determining cluster distances.)
3. Describe the distribution of the three types of clusters in our galaxy. (See p. 487.)
4. Discuss the origin of clusters. (See p. 495.)

Word Practice

*1. The nearest cluster is the _____ _____ cluster.

*2. Interstellar _____ makes determination of cluster distances more difficult, but the associated reddening can be used to make a correction for this effect.

3. The _____ of open clusters occurs because of fast-moving stars and tidal forces.

4. _____ are often larger than open clusters and are rich in young stars.

5. The best method to determine cluster distances is to use _____ _____ stars.

6. _____ clusters are moderately close-knit, irregularly shaped groupings of stars.

*7. Star _____ can be used to determine cluster distances by fitting the cluster to an H-R diagram.

*8. The _____ of globular clusters show that they formed with very little angular momentum.

9. The _____ of globular clusters are all about 12.5 billion years.

*10. Cluster _____ can be used to estimate cluster distances by applying the small angle equation.

11. _____ clusters are massive, spherically symmetric star systems.

173

12. Some globular clusters are sources of _____ of irregular variation.

13. _____ _____ have prominent T Tauri variables.

14. _____ _____ have prominent O and B stars.

*15. _____, the most basic method of star distance measurement, are almost useless for clusters because most clusters are too far away.

*16. The _____ are an example of a naked-eye, open cluster.

17. The _____ _____ of a cluster can be used to estimate the age of the cluster.

18. _____ are basic groupings of stars of probable common origin.

19. The _____ of clusters of various masses depends on the density and temperature of the original diffuse medium.

*20. _____ is a globular cluster located about 13,000 parsecs from us.

Answers to Word Practice

1. Ursa Major		11. globular	
2. obscuration		12. X-rays	
3. disruption		13. T associations	
4. associations		14. O associations	
5. Cepheid variable		15. parallaxes	
6. open		16. Pleiades	
7. luminosities		17. H-R diagram	
8. dynamics		18. clusters	
9. ages		19. origin	
10. diameters		20. M3	

Chapter 23
The Milky Way Galaxy

Answers to Problems in the Text

1. Take recorded history as 3000–4000 years. Use 50 years for the time we have known about our galaxy. Calculate
$$\frac{3500 - 50}{3500} \times 100 = 98.6\%.$$

2. The Milky Way appears as a relatively thin region lying on a great circle across the sky, although we are slightly out of the plane and can see part of one side of the nuclear bulge. That is, it cuts the celestial sphere in roughly two equal parts. If we were well above the plane of the Milky Way, the division of the sky would be uneven, with more sky above the Milky Way visible on the side on which we lie. Also, if we were above, we would be able to see the spiral arms and the disk more clearly.

3. The swarm of open clusters occupies a disk, whereas the globular clusters are distributed in a sphere. The open clusters are Population I, which is younger, whereas the Population II, older, stars are distributed in a more spherical volume. This implies that early star formation occurred before the collapse to a disk.

4. Calculate the center and size of the sphere that the globular clusters define. The size is the size of the galaxy. We can also use radar astronomy to measure the distances of HI clouds.

5. Assume that globular clusters are spherical in distribution. Calculate the distance and direction of this sphere's center. This is also consistent with the direction obtained from 21-cm hydrogen (HI) radiation.

6. First, we see spiral structure in certain other galaxies. Second, if we map open clusters, associations, nebulae, and so on, they fall into bands that can be interpreted as parts of spiral arms. Finally, the 21-cm radiation traces out details of the rest of the arm structure. The following objects reveal strong spiral structure: a, c, d, and f.

7. No. The stars' proper motions will cause them to appear to distort. Most constellations are composed of more than one cluster, so the changes will be extreme. Also, the stars will have gone halfway around the galaxy. Thus, we expect quite a bit of mixing and different neighbors.

8. From the point of view of the rest of the universe, it is the spiral-arm stars that are revolving about the galactic center. Although individual Population II stars are revolving about the galactic center, they have mixed orbits of all inclinations. The net sum of all this motion is no rotation with respect to external galaxies. High-velocity stars are halo stars. So it is the spiral arms that have systematic high velocity relative to the external galaxies.

9. Only old stars (Population II) are in the galactic halo. All original O and B stars have evolved off the main sequence. No new stars are forming there.

10. A 3-kiloparsec arm that is an expanding hydrogen cloud moving outward at 53 km/s; strong infrared and radio radiation from the center; our galaxy is similar to other galaxies that have bright, violent nuclei.

11. Distance to galactic center = 9 kiloparsecs = 9×10^3 parsec = D;

   $\alpha'' = 1''$, $d = (\alpha'' D)/206,265$

   $d = 9 \times 10^3/(2.06 \times 10^5)$ parsec = 4.4×10^{-2} parsec = 0.044 parsec

   a. No, dust blocks off visible light.

   b. Yes, if it is a radio source since radio waves pass through galactic dust. However, there would be no Doppler shift from which to get distances.

12. Convert 230 km/s to parsecs/year.

   $$V = 230 \text{ km/s} \times \frac{\text{parsec}}{3 \times 10^{18} \text{ cm}} \times \frac{10^5 \text{ cm}}{\text{km}} \times \frac{\pi \times 10^7 \text{ s}}{\text{year}}$$

   $$= \frac{2.3 \times \pi \times 10^{14}}{3 \times 10^{18}} \text{ parsec/year}$$

   $$= 2.4 \times 10^{-4} \text{ parsecs/year}$$

   $$\text{period} = \frac{\text{circumference}}{V} = \frac{2\pi \times 9 \times 10^3 \text{ pc}}{2.4 \times 10^{-4} \text{ pc/year}}$$

   $$= 2.3 \times 10^8 \text{ years} = 230 \text{ million years}$$

13. $V_{\text{circ}} = \sqrt{GM/R} \rightarrow M = RV^2/G$; $G = 6.67 \times 10^{-11}$

   $R = 9 \times 10^3 \times 3 \times 10^{16} \text{ m} = 2.7 \times 10^{20} \text{ m}$

   $V^2 = (2.3 \times 10^5 \text{m/s})^2 = 5.3 \times 10^{10} \text{ m}^2/\text{s}^2$

   $$M = \frac{2.7 \times 5.3 \times 10^{30}}{6.67 \times 10^{-11}} = 2 \times 10^{41} \text{ kg}$$

   Two reasons why incorrect: (1) Value of r is only approximately known. (2) We made the approximation that all the mass of the galaxy is at the center.

14. Student response: Diameter 30,000 parsecs = 30,000 parsecs × 3.26 light-years/parsec = 100,000 light-years. In 70 years, it could go 70 light-years.

   $$\% = \frac{70}{10^5} \times 100 = 0.07\%$$

   This result is incorrect since rapidly moving clocks slow down; however, students do not have the background from this course to correctly calculate the result (at light speed, no time elapses). The results given in this and the next problem tell us how much time would pass on Earth for those who stayed behind.

15. Student response: The Orion nebula is 1600 years away at the speed of light. At 20 years per generation,

   $$\text{number of generations} = \frac{1600}{20} = 80.$$

   Result incorrect: See comment after problem 14.

16. This problem is only an approximation since large relativistic effects would occur.

   a. $v = (GM/R)^{1/2}$, $M = 10^5$ solar masses, $R = 10$ AU

   $v = 100 \times$ velocity of Earth in orbit

   Velocity of Earth in orbit = circumference/time

   $$= 2 \times \pi \times 1.5 \times 10^{11} \text{ m}/(\pi \times 10^7 \text{ s})$$

   $$= 3 \times 10^4 \text{ m/s}$$

   so $v = 3 \times 10^6 \text{ m/s}$.

b. $(1/2)mv^2 = (3/2)kT$, so $T = mv^2/(3k)$

Assume hydrogen, so $m = 1.67 \times 10^{-27}$ kg and $k = 1.38 \times 10^{-23}$ in SI. This gives $T = 3.6 \times 10^8$ K.

c. $L = 0.00290/T = 8 \times 10^{-12}$ m = 8 pm

d. This is too short a wavelength to observe on the Earth's surface since the atmosphere is opaque. We need a space-based gamma ray detector.

Sample Test Questions

True-False

1. The first astronomer to confirm that the Milky Way was made of stars was Galileo. T
2. Herschel, in the late 1700s and early 1800s, counted many stars in the Milky Way. T
3. The first astronomer to put the Sun out toward one edge of the disk of the galaxy was H. Shapley. T
4. The overall diameter of the galaxy is 9000 parsecs. F
5. The disk shape of the spiral arms shows that this part of the galaxy has no net angular momentum. F
6. The spiral-arm structure of our galaxy is indicated by the distribution of O and B stars. T
7. Radio waves do not pass as easily through dust as visible light does. F
8. HII regions are most useful for mapping the distant parts of the galaxy's structure. F
9. The galaxy has spiral arms because matter is being jetted out at the center of the galaxy. F
10. The spiral arms can be explained by the density-wave theory. T
11. There are about 2×10^{11} stars in the galaxy. T
12. The mass of the galaxy is about 2×10^{30} kg. F
13. Population I stars are older than Population II stars. F
14. Population II stars have less-heavy elements than Population I stars. T
15. Population I Cepheids are the same as Population II Cepheids. F
16. The 3-kiloparsec arm is between us and the galactic center. T

Multiple Choice

1. It is believed that the globular clusters are distributed spherically with a central concentration in the vicinity of the galactic center.
   *A. correct
   B. wrong: This describes the stars on spiral arms.
   C. wrong: This describes associations.
   D. wrong: They are disk-shaped with spiral arms, not spherical.
   E. wrong: They are centered about the Sun.

2. The Sun is _____ kiloparsecs from the galactic center.
   A. 2–3
   B. 4–6
   C. 6–8
   *D. 8–10
   E. 10–12

3. The overall diameter of the galaxy is estimated at _____ kiloparsecs.
   A. 6–8
   B. 8–10
   C. 20
   D. 25
   *E. 30

4. The Sun's galactic motion is toward the center of the galaxy.
   A. correct
   B. wrong: The Sun is moving away from the galactic center.
   C. wrong: The Sun's motion is an elongated ellipse.
   *D. wrong: The Sun moves in a circular orbit.
   E. wrong: None of the above is correct.

5. The time it takes the Sun to travel around the galaxy is about
   _____ years.
   A. 230
   B. 2.3×10^5
   *C. 2.3×10^8
   D. 2.3×10^{11}
   E. none of these

6. Spiral-arm stars that are farther from the galactic center than the Sun move more slowly than the Sun, while stars closer to the center move faster. This is called
   A. splitting
   B. rigid body motion
   C. passing motion
   *D. differential rotation
   E. none of these

7. Optically, we can see the parts of _____ arms of our galaxy.
   A. 1
   B. 2
   *C. 3
   D. 4
   E. 5

8. The reference direction for the galactic system is the
   A. vernal equinox
   B. star Vega
   C. north point
   *D. galactic center
   E. none of these

9. Galactic latitude is measured
   *A. perpendicular to the plane of the Milky Way
   B. in the plane of the Milky Way
   C. perpendicular to the celestial equator
   D. by stars' proper motion
   E. none of the above

178

10. The general structure of the spiral arms of our galaxy can be mapped using
    A. visible light
   *B. the 21-cm radiation of neutral hydrogen
    C. the 10-megahertz frequency of WWV
    D. both A and B
    E. all three, A, B, and C

Use the following diagram to answer questions 11–12.

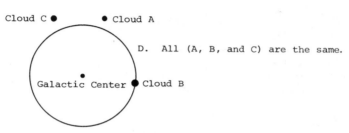

11. Which cloud will have the largest space velocity? __B__ .

12. Which cloud will have the largest Doppler shift? __B__ .

13. We are able to see optically most of the spiral arms in our galaxy. These spiral arms
    are believed to be density waves.
    A. correct
   *B. wrong: Optically, we can see only nearby stars in spiral arms.
    C. wrong: The spiral arms are permanent in the sense that a star in the middle of a
       spiral arm remains in the middle of that arm.
    D. wrong: Both B and C are needed.
    E. misleading: The galaxy has no spiral arms; it is a globular galaxy.

14. To understand why the spiral arms are not wound tighter than they are, it seems
    necessary to assume
    A. that the spiral arms are very young
    B. misleading: We are wrong about how tightly the spiral arms are wound.
   *C. that the spiral arms are density waves
    D. both A and C
    E. none of the above

15. The mass of the galaxy is about _____ kilograms.
    A. 2×10^{30} *D. 4×10^{41}
    B. 3×10^{33} E. 7×10^{69}
    C. 2×10^{44}

179

16. There are about _____ stars in the Milky Way galaxy.
   A. 3×10^{10}
   *B. 2×10^{11}
   C. 2×10^{44}
   D. 9×10^{57}
   E. Astronomers have no idea.

17. Which of the following are composed of Population I stars?
   A. globular clusters
   *B. spiral arms
   C. the corona
   D. both B and C
   E. all three, A, B, and C

18. Which of the following is not evidence for two types (or populations) of stars?
   A. stellar chemistry
   B. ages
   C. motions and orbits
   D. different types of Cepheid stars
   *E. All those listed are evidence for two populations.

19. Which are the older stars in the galaxy?
   A. Population I
   *B. Population II
   C. Population A
   D. Population B
   E. Population III

20. Which stars have virtually no elements heavier than hydrogen and helium?
   A. Population I
   *B. Population II
   C. Population A
   D. Population B
   E. Population III

21. The point in the sky toward which the Sun appears to be moving is the
   *A. solar apex
   B. local standard of rest
   C. peculiar velocity
   D. motion point
   E. none of these

22. The galactic nucleus had a large energy release about _____ years ago, as shown by the 3-kiloparsec arm.
   A. 10 million
   B. 50 million
   *C. 100 million
   D. 1 billion
   E. 12 billion

23. Which of the following is not proposed as a condition that could have released large amounts of energy in the galactic center?
   A. unstable stars of very large mass undergoing nova
   B. stellar collisions
   *C. many K and M stars going through their life cycle
   D. dust falling on high-density, high-gravity objects
   E. All of the above are possible mechanisms.

Problem Involving Optional Mathematical Equations from Another Chapter

*24. (Optional Equation VII) A high-velocity star is approaching the Sun at 120 km/s.
At what wavelength (approximately) would a 479.6 nm line be observed?

   A. 477.6 nm   D. 479.8 nm
*B. 479.4 nm   E. 481.6 nm
   C. 479.6 nm

Essay

1. Describe some of the problems astronomers had to overcome to discover the structure of our galaxy. (See pp. 498–499.)
2. Describe how radio astronomy is able to map the spiral-arm structure. (See pp. 506–507.)
3. Explain how the density-wave theory works. (See p. 507.)
4. Describe the modified coffee cup model (chain reaction) of the spiral arms. (See pp. 507–508.)
5. Discuss three of the types of evidence for two stellar populations. (See Table 23-2.)
6. What are the three pieces of evidence for high-energy events in the galactic nucleus, and what are three possible mechanisms to cause these events? (See pp. 516–517.)
7. Summarize the evolution of the galaxy. (See p. 519.)

Word Practice

*1. The spiral arms are contained in the _____ of the galaxy.

2. The galactic _____ is a spherical swarm of stars and globular clusters.

3. The _____ of our galaxy is its center.

4. _____ stars belong to the galactic halo and are usually Population II.

5. _____ _____ is the coordinate from the galactic system that is measured in the plane of the Milky Way.

6. The _____ _____ of our galaxy is best investigated with radio telescopes.

7. _____ _____ _____ _____ is an imaginary point in the middle of our local swarm moving in a circular orbit around the galactic center.

*8. Two periodic motions may combine to produce the _____ that is the spiral arms.

9. The _____ _____ stars are the younger stars in our galaxy.

10. _____ is the abbreviation for local standard of rest.

11. Star formation is most rapid in the spiral _____ of our galaxy.

12. Galactic _____ is measured perpendicular to the Milky Way.

13. The _____ _____ _____ _____ is about 12 billion years, so the spiral arms must be patterns.

14. The galaxy has a(n) _____ of about 30,000 parsecs.

15. The _____ velocity of a star is its velocity relative to the LSR.

*16. The spiral arm _____ of the galaxy rotate about the nucleus in approximately circular orbits with a velocity that depends upon their distance from the galactic center.

17. The _____ theory explains how the spiral arms can last so long.

18. The _____ _____ is the point in the sky toward which the Sun appears to move.

19. Both Cepheid II and _____ _____ _____ are Population II stars.

20. One thousand parsecs equal one _____.

21. The period of _____ of the Sun about the galactic center is 230 million years.

*22. _____ regions can emit 21-cm radiation.

23. _____ rotation of the galaxy means that the period of rotation of a disk star depends upon its distance from the galactic center.

24. The _____ of the galaxy is about 10^{11} M_0.

25. The _____ of the Population II stars more closely reflects the percentages of elements in the early galaxy than that of Population I stars.

26. The two types of _____ have different period-luminosity curves.

27. _____ _____ _____ are used to map the spiral arm structure of our galaxy.

28. The _____ _____ runs along the center of the Milky Way.

29. The _____ _____ stars comprise the older population of the galaxy.

*30. _____ telescopes are the most useful for studying the spiral arm structure of our galaxy.

*31. The _____ of globular cluster stars differs from that of the disk stars.

32. Our own galaxy is called the _____ _____ galaxy.

Answers to Word Practice

1. disk
2. halo
3. nucleus
4. high-velocity
5. galactic longitude
6. spiral (or galactic) structure
7. local standard of rest
8. pattern
9. Population I
10. LSR
11. arms
12. latitude
13. age of the galaxy
14. dimension (or diameter)
15. peculiar
16. stars
17. density-wave
18. solar apex
19. RR Lyrae stars
20. kiloparsec
21. revolution
22. HI
23. differential
24. mass
25. composition
26. Cepheids
27. 21-cm radio waves
28. galactic equator
29. Population II
30. radio
31. chemistry (or composition)
32. Milky Way

Chapter 24
The Local Galaxies

Answers to Problems in the Text

1. Spiral. Spiral and giant elliptical. Elliptical. (The largest and brightest galaxies, although rare, are probably giant ellipticals.)
2. Ordinary elliptical.
3. 670 kiloparsecs $= 670 \times 3.26$ kilolight-years
   $= 2.18 \times 10^6$ light-years
   $=$ about 2 million years
4. The one with the K spectrum would probably be an elliptical (predominant red giants); the one with B and A could not be an elliptical and would presumably be a spiral galaxy (Population I).
5. Galaxy size = 30 kiloparsecs; distance from Milky Way to Andromeda galaxy = 670 kiloparsecs, so separation is 670/30 ~ 24 disk diameters.
6. No. The local group has a common center of mass (barycenter). If there is a net rotation about this center, it would define an axis of rotation, but this would be hard to detect and has not been found. Distant galaxies recede from each other and appear to be radially receding from us also so that no axis of symmetry or rotation is known. Commonly, astronomers use the Milky Way galaxy.
7. They are too faint to be observed at those distances.
8. No. This only shows that star formation is going on at a rapid rate in these galaxies. The galaxies themselves may be very old.
9. Hot blue stars are young, the only other stars as bright as red supergiants, which are old stars. Young red stars are very faint. Star-forming regions would be expected to have young stars, the brightest of which are blue O and B.
   a. O and B stars evolve quickly. When star formation is slow, there is no longer a large source of Os and Bs, so they are not seen.
   b. Population I.

Advanced Problems

10. a. $\dfrac{\alpha"}{206,265} = \dfrac{d}{D} = \dfrac{40}{670}$

   $\alpha" = \dfrac{40 \times 2.06 \times 10^5}{670} = 1.2 \times 10^4$ seconds of arc $3°$

   This is six times the angular size of the Moon! But most of it is too faint to be seen. Consider the sky's appearance if it were not so faint.

   b. $\alpha" = \dfrac{d}{D} \times 2.06 \times 10^5 = \dfrac{3 \times 10^{-3}}{670} \times 2.06 \times 10^5 = 0."9$

   This is at the limit of resolution. (See Fig. 24-14, p. 533.)

1. $\dfrac{\text{change in wavelength}}{\text{wavelength}} = \dfrac{v}{c} = \dfrac{0.165}{500} = 3.3 \times 10^{-4}$

Therefore, $v = 3.3 \times 10^{-4}c = 99$ km/s.

2. $V_{circ} = \sqrt{GM/R} \rightarrow M_{\odot}\dfrac{RV^2}{G}$

$$M = \dfrac{4 \times 10^3 \times 3 \times 10^{16} \times [9.9 \times (10^4)]^2}{6.67 \times 10^{-11}}$$

$$= 1.8 \times 10^{40} \text{ kg or } 10^{10}\ M_{\odot}$$

This is enough mass for 10 billion Suns or about 1/10 of our galaxy's mass.

Sample Test Questions

True-False

1. A galaxy's age depends upon its location on the "tuning fork" diagram. F
2. The Doppler shift can be used to determine the rotation of some other galaxies. T
3. The Magellanic clouds are satellite galaxies of the Milky Way. T
4. The Magellanic clouds are regular galaxies. F
5. The Tarantula nebula (30 Doradus) is believed to be the partially formed nucleus of the Large Magellanic cloud. T
6. The Magellanic clouds contain only Population I stars. F
7. Galaxies resembling our galaxy but without the spiral arms and dust are called elliptical galaxies. T
8. Galaxies tend to cluster in different-size groups. T
9. All elliptical galaxies are relatively small galaxies. F
10. The classification system for galaxies is often referred to as the "tuning fork" diagram. T
11. Approximately 70% of all galaxies are believed to be spirals. F

Multiple Choice

1. The mass of a galaxy can sometimes be obtained from
   A. Kepler's laws and the rotation speed of the galaxy
   B. the motion of binary galaxies
   C. Wien's law
   *D. both A and B
   E. none of the above

2. Which of the following nearby galaxies is a spiral galaxy?
   A. Large Magellanic cloud   D. Maffei I
   B. Draco system   *E. both C and D
   C. Andromeda galaxy

3. Which of the following nearby galaxies is an elliptical galaxy?
   A. Large Magellanic cloud   D. Maffei I
   *B. Draco system   E. both C and D
   C. Andromeda galaxy

4. Which of the following nearby galaxies is an irregular galaxy?
   *A. Large Magellanic cloud
   B. Draco system
   C. Andromeda galaxy
   D. Maffei I
   E. both C and D

5. The Magellanic clouds are _____ galaxies.
   A. regular
   *B. irregular
   C. spiral
   D. elliptical
   E. intergalactic globular cluster

6. The chief importance of the Magellanic clouds to modern astronomy is that they
   A. are irregular galaxies
   B. have many O and B stars
   C. show that galaxies can have satellite galaxies
   *D. provide us with a collection of objects all at effectively the same distance
   E. none of the above

7. Which galaxies are dominated by Population II stars and have little gas or dust?
   A. spiral
   B. irregular
   C. elliptical
   D. intergalactic globular
   *E. both C and D

8. In the Andromeda galaxy, star formation seems to be occurring most rapidly in the
   A. nucleus
   B. inner arms
   *C. intermediate and outer arms
   D. halo
   E. misleading: Little or no star formation is occurring.

9. The galaxies near us (say out to 1000 kiloparsecs) are called
   *A. the Local Group
   B. the nearby crowd
   C. the near ones
   D. M33
   E. none of these

10. A spiral galaxy whose central region is a bright, barlike object is called a
    _____ galaxy.
    A. bright, barlike
    *B. barred spiral
    C. giant elliptical
    D. barred irregular
    E. none of these

11. The classification scheme for galaxies is called Hubble's
    A. classification scheme
    *B. tuning fork diagram
    C. two-pronged classification
    D. diagram
    E. relation

The following diagrams are for questions 12–16.

12. Which galaxy is an EO galaxy? __A__ .

13. Which galaxy is an SO galaxy? __D__ .

14. Which galaxy is an Sc galaxy? __E__ .

15. Which galaxy is an Sa galaxy? __B__ .

16. Which galaxy is an SBb galaxy? __C__ .

17. Galaxies with amorphous appearance, no sharp nucleus, and some gas and dust and Population II stars are called
    A. elliptical
    B. spiral
    C. intergalactic globular cluster
    D. irregular I
    *E. irregular II

18. Elliptical galaxies comprise about _____% of all galaxies (counting intergalactic globular elliptical).
    *A. 70   D. 15
    B. 50   E. 3
    C. 25

19. Spiral galaxies comprise about _____% of all galaxies.
    A. 70   *D. 15
    B. 50   E. 3
    C. 25

20. A galaxy's place in the "Hubble's tuning fork" is probably determined by its
    A. age
    *B. initial conditions of mass, rotation, turbulence, and so on
    C. mass
    D. both A and B
    E. both A and C

Essay

1. Describe the differences and similarities of spiral, elliptical, and irregular galaxies.
2. Describe the Hubble tuning fork classification scheme. Draw the relevant diagram. (See pp. 531–541 and Fig. 24-25.)
3. Describe the Local Group: types of members, distribution in space, and so on. (See Table 24-1 and Fig. 24-1.)

187

4. Summarize how our theory of stellar evolution is successful in describing how similar galaxies of great (equal?) age could have widely different populations of stars (See pp. 547–550.)
5. Discuss the importance of the Andromeda galaxy in stellar evolution theory. (See p. 531.)

Word Practice

*1. The differences in galaxies probably are due solely to different _____ _____, not to evolutionary effects of ages.

2. The _____ of galaxies can be determined by measuring Doppler shifts and applying Kepler's laws.

3. _____ _____ galaxies are spiral galaxies whose central regions are bright and barlike.

4. _____ _____ stars are the older stars in the galaxy.

5. _____ _____ stars are the younger stars in the galaxy.

6. Our galaxy is a member of a cluster of galaxies called the _____ _____.

7. The Magellanic _____ is a bridge of hydrogen that may connect the two Magellanic clouds to our galaxy.

8. _____ galaxies show little or no structure.

9. Our galaxy is a(n) _____ galaxy.

*10. Both Magellanic clouds contain a few _____ _____ _____ unlike the red-giant-dominated clusters of our galaxy.

11. _____ galaxies resemble giant globular clusters.

*12. _____ shifts from either side of a galaxy's center allow us to determine the rotation of the galaxy.

13. The _____ _____ measures the amount of mass required to produce the same amount of stellar radiation. It is high in ellipticals.

14. _____ _____ clusters may be small examples of elliptical galaxies.

15. The _____ galaxy is the nearest spiral galaxy to the Milky Way.

16. The _____ _____ is another name for 30 Doradus.

17. _____ galaxies are the most common type in our group.

*18. By measuring the Doppler shifts on either side of a galaxy's center and using

_____ _____, we can determine the galaxy's mass.

19. The _____ galaxy is similar to our galaxy except that it is more open and has blue globular clusters.

20. The _____ _____ of a galaxy, or region of a galaxy, will depend upon the material left for star formation.

21. _____ _____ is believed to be the partially formed nucleus of the Large Magellanic cloud.

22. The _____ _____ are the two satellite galaxies of the Milky Way.

23. A(n) _____ _____ galaxy is like our galaxy except that the nucleus is bar-shaped.

24. _____ galaxies are like our galaxy without the disk and spiral arm structure.

Answers to Word Practice

1. initial conditions
2. masses
3. barred spiral
4. Population II
5. Population I
6. Local Group
7. stream
8. irregular
9. spiral
10. blue globular clusters
11. elliptical
12. Doppler
13. mass/luminosity ratio
14. intergalactic globular
15. Andromeda
16. Tarantula nebula
17. elliptical
18. Kepler's laws
19. Triangulum
20. stellar population
21. 30 Doradus
22. Magellanic clouds
23. barred spiral
24. elliptical

Chapter 25
The Expanding Universe of Distant Galaxies

Answers to Problems in the Text

1. 1 parsec $= 3 \times 10^{16}$ m

   1 megaparsec $= 10^6$ parsecs $= 3 \times 10^{22}$ m

   1 mile $= 1.609$ km $= 1.609 \times 10^3$ m

   $$1 \text{ megaparsec} = \frac{3 \times 10^{22} \text{ m}}{1.609 \times 10^3 \text{ m/mile}} = 1.9 \times 10^{19} \text{ miles}$$

   The teacher must demonstrate how the dimensions must check out and must show how this helps in setting up the problem; that is, show that "m" cancels "m," leaving miles.

2. a. Red shift.

   b. Velocity is 6700 km/s, receding. The red shift could be expressed as

   $$\frac{v}{c} = \frac{6700}{3 \times 10^5} = 2.2 \times 10^{-2}, \text{ that is, a shift of 2.2\%.}$$

   c. They would reach the same conclusion about us that we have reached about them: We are receding from them at 6700 km/s.

3. They have small, bright centers that produce large, varying amounts of energy. A Seyfert nucleus, seen from too far away for the arms to be resolved, would resemble a weak quasar.

4. These imply that red shift is not entirely cosmological, so there may be another explanation for the red shift of quasars. If there were another explanation, we could not determine quasar distances from their red shifts.

5. Our galaxy is a relatively weak radio emitter. Radio galaxies will emit hundreds to millions more units of energy at radio wavelengths.

6. The farther out we look, the farther into the past we see. We see distant galaxies as they were many billions of years ago since they are billions of light-years away. Some of the most distant are more than 10 billion light-years away, so their light shows them as they were when our galaxy was just forming.

Advanced Problems

7. Normal automobile speeds are less than 100 km/hour, that is, less than 28 m/s; hence,

   $$\frac{v}{c} = \frac{28 \text{ m/s}}{3 \times 10^8 \text{ m/s}} = 10^{-7}.$$

   Thus, the wavelength shift is only 1 part in 10 million—too small to detect since the band of frequency covered by a single radio station is very wide compared to this.

8. a. Shift $= 3.4 \times$ wavelength $= 3.4 \times 500$ nm $= 1700$ nm

   $= 1.7$ microns $= 1.7$ μm.

   b. Fainter.

   c. It would be extremely hard to detect. Visible radiation would be very faint since all visible light is shifted to red. Apparent temperature is very cool:

$$T = \frac{0.0029}{17 \times 10^{-3}} = 1700 \text{ K}.$$

Also, it would not be as intrinsically bright in radio as a quasar, so we would not find radio telescopes helpful.

9. $V = 5700$ km/s; $V = Hr \rightarrow r = V/H$

If $H = 57$ km/s/Mpc,

$$r = \frac{5700 \text{ km/s}}{57 \text{ km/s/Mpc}} = 100 \text{ Mpc}.$$

If $H = 100$ km/s/Mpc,

$$r = \frac{5700 \text{ km/s}}{100 \text{ km/s/Mpc}} = 57 \text{ Mpc}.$$

10. $\alpha'' = (d/D) \times 2.06 \times 10^5$

If $H = 57$,

$$\alpha'' = \frac{30 \times 10^3 \text{ pc}}{10^2 \times 10^6 \text{ pc}} \times 2.06 \times 10^5 = 60'' = 1'.$$

If $H = 100$,

$\alpha'' = 105'' = 1'.8$.

11. a. Use $\Delta\lambda/\lambda = v/c$ for classical

and $\Delta\lambda/\lambda = [(1 + v/c)/(1 - v/c)]^{1/2} - 1$ for relativistic.

v/c	Classical	Relativistic
.01	.01	.01005
.1	.1	.106
.5	.5	.73
.9	.9	3.36
.99	.99	13.1

b. At about .1. Define error as

error $= [(\Delta\lambda/\lambda)_{rel} - (\Delta\lambda/\lambda)_{class}]/(\Delta\lambda/\lambda)_{class}$.

At $v/c = .1$, the error is 0.06 or 6%.

12. From 11a, it is at about .9. More exactly,

$3.78 = [(1 + v/c)/(1 - v/c)]^{1/2} - 1$,

so $(1 + v/c)/(1 - v/c) = (4.78 \times 4.78) = 22.85$,

which gives $v/c = 0.916$.

Sample Test Questions

True-False

1. Some astronomers have seriously suggested that even clusters of galaxies cluster. T
2. All galaxies are red-shifted. F
3. All distant clusters of galaxies are red-shifted. T
4. The red shift in a cluster of galaxies is found to be directly proportional to the distance when this distance can be determined by other means. T

5. The relationship between velocity of recession and distance for clusters of galaxies is called Hubble's relation. T
6. Since distant galaxies are receding from us, we are at the center of the universe. F
7. The only possible interpretation of the red shift of galaxies is that it is a Doppler shift. F
8. At least some observed red shifts may be gravitational red shifts. T
9. Our galaxy is a radio galaxy. F
10. The radio radiation of most radio galaxies is synchrotron radiation. T
11. There is no evidence that galaxies ever collide. F
12. *Quasar* stands for quasi-stellar radium source. F
13. Many quasars have large red shifts. T
14. If the red shift of quasars is cosmological, we have no difficulty explaining their energy generation. F
15. If a source brightens and dims regularly with a period of one day, its size must not be larger than one light-day. T
16. One possible source of the red shift of quasars is a strong gravitational field. T

Multiple Choice

1. 1 megaparsec = _____ parsecs.
   A. 100
   B. 1000
   *C. 10^6
   D. 10^9
   E. 10^{21}

2. If r represents the distance to a nearby cluster of galaxies and V represents the velocity of recession of the cluster, Hubble's relation states that $r = HV$, where H represents Hubble's constant.
   A. correct
   B. wrong: applies to individual galaxies, not clusters
   *C. wrong: relation is $V = Hr$
   D. wrong: both B and C
   E. wrong: but none of the above

*3. If a cluster of galaxies is receding from us at 15,000 km/s, its distance, by the Hubble relation, is (approximately) _____ megaparsecs. (Take $H = 57$ km/s/Mpc.)
   A. 26
   B. 200
   *C. 260
   D. 960
   E. 2600

4. The cosmological theory of red shifts proposes that the red shifts of galaxies are due to
   *A. Doppler shifts
   B. gravitational forces
   C. photons losing energy
   D. both A and B or either
   E. both B and C or either

192

5. The noncosmological theory of red shifts proposes that the red shifts of galaxies are due to
   A. Doppler shifts
   B. gravitational forces
   C. photons losing energy
   D. both A and B or either
   *E. both B and C or either

6. The radio-emitting regions of most optically visible radio galaxies are often
   A. larger than the optically visible galaxy
   B. concentrated in two lobes in either side of the galaxy
   C. associated with ordinary spiral galaxies
   *D. both A and B
   E. all three, A, B, and C

7. When two galaxies collide
   A. many stellar collisions occur, releasing energy
   *B. their gas clouds collide, emitting radio waves
   C. their total angular momentum is changed
   D. both A and B
   E. all three, A, B, and C

8. A Seyfert galaxy
   *A. is a spiral galaxy with a small, bright region in its nucleus that shows broad, bright emission lines
   B. is an elliptical galaxy with a small, bright, nearly stellar-appearing nucleus
   C. has a relatively small size and high surface brightness
   D. is a supergiant elliptical found near the center of a cluster of galaxies
   E. is a highly irregular galaxy, completely amorphous, without any resolved stars

9. Quasars appear as
   A. spiral galaxies with extended nuclei
   *B. strong radio sources of small sizes that are sometimes optically identified with a starlike object (unresolved) and often have large, red-shifted spectral lines
   C. strong ultraviolet and infrared sources
   D. X-ray sources
   E. spherical globes, reddish in color

10. Other than a Doppler shift, what is the next most popular interpretation of the cause of red shift in quasars?
   A. rotational   D. magnetic
   B. temperature   E. no other interpretation
   *C. gravitational

11. Some quasars seem to be related to _____ galaxies.
    A. elliptical
    B. irregular
    *C. Seyfert
    D. gumball
    E. intergalactic globular clusters

12. A quasar radio signal varies by a factor of 5, returning to its original brightness about every two months. Hence, the radio-emitting region is no more than _____ in size.
    A. 1/2 light-month (1.2×10^{-2} megaparsecs)
    B. 1 light-month (2.5×10^{-2} megaparsecs)
    *C. 2 light-months (5×10^{-2} megaparsecs)
    D. 1 light-year (0.3 megaparsec)
    E. not near any of these

13. The cosmological interpretation of quasars gives which of the following properties to quasars with large red shifts?
    A. great distance
    B. relative proximity
    C. large amount of energy
    *D. both A and C
    E. both B and C

14. The noncosmological interpretation of quasars gives which of the following properties to quasars with large red shifts?
    A. great distance
    *B. relative proximity
    C. large amounts of energy
    D. both A and C
    E. both B and C

15. If quasars are cosmological, which of the following might explain their energy generation?
    A. supersupernovae of supermassive superstars
    B. novae
    C. rapidly rotating massive objects with strong magnetic fields
    *D. both A and C
    E. all three, A, B, and C

*16. What is the velocity of recession as a fraction of the speed of light?
    A. 0.5
    *B. 0.8
    C. 0.9
    D. 0.99
    E. not enough information to tell

Essay

1. Describe two possible causes of the red shift of distant galaxies. (See pp. 558–559.)
2. What are some of the mysteries that still confront us about galactic nuclei? (See pp. 563–568.)

3. What are the two interpretations of quasars given in the text? (See pp. 574–578.)
4. Compare how much we know about galaxies and remote galaxylike objects with how much we know about planets in our solar system.

Word Practice

1. The possible associations of clusters of galaxies are called _____.

2. A galaxy may appear distorted by _____ _____ if another, closer galaxy lies nearby in the same direction.

3. The theory of _____ _____ _____ states that red shifts are due to the Doppler effect of recession.

4. _____ are radio sources that often have large red shifts. If the red shifts are cosmological, many of these are the brightest objects known.

5. Hubble's _____ is a proportionality between distance and recession velocity for clusters of galaxies.

6. The cosmological interpretation of red shift has the universe _____.

7. Strong gravitational fields can cause _____ _____ _____.

8. _____ _____ is produced by a hot gas in which ions in magnetic fields are accelerated close to the speed of light.

9. Galaxies with evidence of explosions in their nuclei or with brighter than normal nuclei are called galaxies with _____ _____.

10. The _____ _____ of clusters of galaxies is such that we cannot say that any cluster is at the center of the universe.

11. One million parsecs are one _____.

12. Radio-quiet quasars are called _____.

13. The theory of cosmological red shifts interprets the shifts as _____ _____.

14. An example of one of the theories of _____ _____ _____ is photons losing energy over long distances.

15. _____ _____ relates the recession velocity of a cluster of galaxies to its distance from us.

16. _____ _____ are galaxies with bright centers.

195

17. Galaxies that are strong radio emitters are called _____ _____.

18. Observationally, most galaxies show a(n) _____ _____ in their spectral lines.

Answers to Word Practice

1. superclusters
2. gravitational lensing
3. cosmological red shifts
4. quasars
5. constant
6. expanding
7. gravitational red shifts
8. synchrotron radiation
9. active nuclei
10. mutual recession
11. megaparsec
12. QSOs
13. Doppler shifts
14. noncosmological red shifts
15. Hubble's relation
16. Seyfert galaxies
17. radio galaxies
18. red shift

Chapter 26
Cosmology: The Universe's Structure

Answers to Problems in the Text

1. There is only one universe to observe; thus, it may not be susceptible to some kinds of scientific analysis.
2. Such a telescope could show us detailed structures of nearby galactic nuclei and of distant galaxies, the shapes of quasars, and so on. It could also help us tell which of the curves in Fig. 26-4 describes the universe.
3. The universe is expanding. Olbers' paradox can (or must) be explained, which the Newtonian model cannot do. The universe has clearly evolved—witness more quasars at large red shifts (z) than at small z.
4. The universe is expanding, so light from distant stars is red-shifted, and we do not believe the universe is old enough to be in thermal equilibrium.
5. No. When we look across large distances, we are looking back in time, so we see the universe at an earlier epoch. In fact, this quasar count is taken to support the big bang cosmologies against the steady state cosmology.
6. a. Cosmological red shifts; excess luminosity of distant galaxies (Fig. 26-9); distribution of quasars; darkness of the night sky.
   b. Distribution of quasars; lack of an explanation of 3-K isotropic black-body radiation.

Advanced Problems

7. $W = 0.00290/T$; W is wavelength, T is temperature.
$$W = \frac{0.00290}{10^6} = 2.9 \times 10^{-9} \text{ m} = 2.9 \text{ nm}$$

a. This wavelength is completely blocked by our atmosphere. A telescope in space would be needed to see it. This wavelength is soft X-ray.

b. This extremely short wavelength is absorbed by the Earth's atmosphere. A space telescope would not have this problem.

c. We want a red shift to change 2.9 nm to 500 nm. That is,
$$\frac{\text{change in wavelength}}{\text{wavelength}} = \frac{500 - 2.9}{2.9} = 171 \gg 1.$$

This is a Doppler factor of $171 = z$, which is much larger than any known. Nor is it likely to be detected because of the 16-billion-year horizon.

d. Use Stefan-Boltzmann law, $E = \sigma A T^4$.
$$\frac{E_*}{E_\odot} = \frac{\sigma A_* T_*^4}{\sigma A_\odot T_\odot^4}$$

We want to compare unit areas of each object.

So, $\left(\dfrac{E_*}{E_\odot}\right)_{\text{unit area}} = \dfrac{T_*^4}{T_\odot^4} = \left(\dfrac{10^6 \text{ K}}{5.7 \times 10^3 \text{ K}}\right)^4 = 5.4 \times 10^6$

A unit surface radiates 5.4×10^6 times as much as the Sun.

8. 91 nm $= 9.1 \times 10^{-10}$ m

$\Delta\lambda/\lambda = [(1 + v/c)/(1 - v/c)]^{1/2} - 1 = 3.36,$

where the last equality follows from Chapter 25, problem 11a. Thus, $\Delta\lambda = 3.36 \times 9.1 \times 10^{-10}$ m $= 3.06 \times 10^{-9}$, which is receding, so add

$\lambda + \Delta\lambda = \lambda_{observed} = 3.97 \times 10^{-9}$ m.

Shifted from the UV to the visible.

Sample Test Questions

True-False

1. The earliest cosmologies (B.C.) made sense of the universe by using mathematics. F
2. The earliest cosmologies (B.C.) could make no testable predictions. T
3. The Newtonian-Euclidean universe was static but evolving. T
4. Olbers' paradox deals with why the sky is blue. F
5. Olbers' paradox is resolved by noting that dust absorbs distant starlight. F
6. The age of the universe is estimated at roughly 16 billion years. T
7. Any valid cosmological theory must be able to explain Olbers' paradox. T
8. Non-Euclidean geometries were developed in the late 1800s and applied to cosmology in 1917. T
9. In any geometry, a three-dimensional space must be infinite. F
10. Universes with infinite volumes are called closed universes. F
11. Static closed-universe models seem not to fit current observational data. T
12. Big bang cosmologies were first developed by G. Lemaître. T
13. All big bang cosmologies propose a definite beginning time for the universe. F
14. Big bang theories are not able to explain why the initial fireball came into existence. T
15. The steady state cosmology assumes that space is the same everywhere and all the time. T
16. The hierarchical universe model allows the universe to have an extremely low density, possibly even approaching zero. T
17. There is some indication that the real universe is uncurved, open, and Euclidean. T
18. According to present data, the universe is closed and finite in volume. F

Multiple Choice

1. A study of the structure of the universe as a single, orderly system is
   A. history
   *B. cosmology
   C. cosmogony
   D. cosmetology
   E. comic

2. The study of the universe's origin is called
   A. history
   B. cosmology
   *C. cosmogony
   D. cosmetology
   E. comic

3. Which of the following correctly sequences distance-measurement techniques in order of increasing distance of applicability?
   A. tape measure, H-R diagram, trigonometric parallax, Cepheids, Hubble relation
   B. tape measure, Hubble relation, H-R diagram, trigonometric parallax, Cepheids
   C. Hubble relation, trigonometric parallax, Cepheids, tape measure, H-R diagram
   *D. tape measure, trigonometric parallax, H-R diagram, Cepheids, Hubble relation
   E. H-R diagram, tape measure, Hubble relation, Cepheids, trigonometric parallax

4. Ancient models of cosmology can be essentially distinguished from others by the phrase
   A. A Euclidean geometry filled with particles obeying Newton's laws
   *B. consideration of the nonmaterial attributes or "essences" of things
   C. a static, curved universe
   D. a universe expanding from a point of maximum density
   E. a universe the same at all times

5. The Newtonian-Euclidean cosmology can be essentially distinguished from others by the phrase
   *A. A Euclidean geometry filled with particles obeying Newton's laws
   B. consideration of the nonmaterial attributes or "essences" of things
   C. a static, curved universe
   D. a universe expanding from a point of maximum density
   E. a universe the same at all times

6. A big bang cosmology can be essentially distinguished from others by the phrase
   A. A Euclidean geometry filled with particles obeying Newton's laws
   B. consideration of the nonmaterial attributes or "essences" of things
   C. a static, curved universe
   *D. a universe expanding from a point of maximum density
   E. a universe the same at all times

7. The steady state cosmology can be essentially distinguished from others by the phrase
   A. A Euclidean geometry filled with particles obeying Newton's laws
   B. consideration of the nonmaterial attributes or "essences" of things
   C. a static, curved universe
   D. a universe expanding from a point of maximum density
   *E. a universe the same at all times

8. Inability to make testable predictions is a limitation of _____ cosmology(ies).
   *A. ancient
   B. Newtonian-Euclidean
   C. big bang
   D. steady state
   E. hierarchical

199

9. Inability to explain Olbers' paradox is a limitation of _____ cosmology(ies).
    A. ancient
    *B. Newtonian-Euclidean
    C. big bang
    D. steady state
    E. hierarchical

10. Inability to explain why the initial fireball came into existence is a limitation of _____ cosmology(ies).
    A. ancient
    B. Newtonian-Euclidean
    *C. big bang
    D. steady state
    E. hierarchical

11. Inability to predict the radiation left from the primeval fireball is a limitation of _____ cosmology(ies).
    A. ancient
    B. Newtonian-Euclidean
    C. big bang
    *D. steady state
    E. hierarchical

12. Inability to show that the superclusters of galaxies are real is a limitation of _____ cosmology(ies).
    A. ancient
    B. Newtonian-Euclidean
    C. big bang
    D. steady state
    *E. hierarchical

13. If the universe is infinite and Newtonian, the sky should be uniformly bright; clearly it is not. The preceding is a statement of
    A. the perfect cosmological principle
    *B. Olbers' paradox
    C. Einstein's postulate
    D. the cosmological principle
    E. Hubble's relation

14. Which diagram of space below, if continued, would be of finite extent? _B_ .

A   B   C

    D. both A and B are of this nature
    E. both A and C are of this nature

15. Which diagram of space above, if continued, would be of infinite extent? _E_ .

5. Which of the following are parts of the cosmological principle?
   A. The universe is homogeneous.
   B. The universe is isotropic.
   C. The universe is the same at all times.
   *D. Both A and B are parts.
   E. All three, A, B, and C, are parts.

7. Which of the following are parts of the cosmological principle used in the steady state cosmology?
   A. The universe is homogeneous.
   B. The universe is isotropic.
   C. The universe is the same at all times
   D. Both A and B are parts.
   *E. All three, A, B, and C, are parts.

8. The statement, "All observers, everywhere in space at one time, would view the same large picture of the universe" is called
   A. the cosmological paradox
   *B. the cosmological principle
   C. the global principle
   D. the perfect cosmological assumption
   E. none of the above

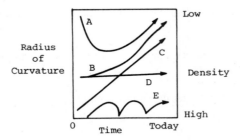

19. In the diagram above, which is a universe that began contracting, reached a minimum, and is now expanding forever? _A_.

20. In the diagram above, which line represents a static universe? _D_.

21. In the diagram above, which line represents a periodically expanding and contracting universe? _E_.

22. We observe that there are fewer radio sources at great distances than we would expect if space were negatively curved.
   A. correct
   B. wrong: We observe more at great distances.
   *C. wrong: We would expect more if space were positively curved.
   D. wrong: We get the number we expect, so space is flat.
   E. wrong: Both B and C are needed.

Use the following diagram for questions 23–26.

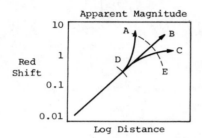

23. In the diagram above, which curve represents the expected red shift versus distance for a closed, spherical universe? __A__.

24. In the diagram above, which curve represents the expected red shift vs. distance for the steady state theory? __C__.

25. In the diagram above, which curve represents the expected red shift vs. distance for an open, Euclidean universe? __B__.

26. In the diagram above, which letter indicates the current limit of observational data? __D__.

Essay

1. Describe the similarities and differences in big bang cosmologies and the steady state theory. (See pp. 591–592.)
2. Explain by analogy how the universe may be expanding, have no edge, and yet be finite. (See pp. 588–589, 592.)
3. Describe the ideas in the hierarchical universe theory and the limitations of the theory. (See p. 592.)
4. Argue for or against the statement, "Modern astronomy has explained the universe better than ancient theologies." (See p. 586.)

Word Practice

1. The _____ comprises all matter and energy in existence anywhere.

2. The _____ _____ cosmology must have new galaxies forming between the old.

*3. _____ _____ is used to find the distance to clusters of galaxies.

*4. _____ stars are used to find the distances to the nearby galaxies.

5. _____ _____ is necessary in the steady state cosmology.

*6. Curved spaces were introduced into cosmology by _____.

7. _____ _____ arises if we assume an infinite Euclidean universe.

*8. A flat space is a(n) _____ space.

9. A(n) _____ _____ has an infinite volume.

10. A(n) _____ universe has finite volume.

11. A(n) _____ universe means that the stars, galaxies, clusters, and so on are uniformly distributed on the large scale.

12. _____ is the study of the structure of the universe.

*13. The distance to the inner planets can be measured by _____.

14. The Newtonian-Euclidean cosmology has a(n) _____ universe.

*15. In the _____ universe model, the galaxies belong to clusters, the clusters to superclusters, and so on.

16. In the _____ _____ _____ , the universe is expanding from a past point of higher density.

17. Euclidean geometry and Newton's laws were used to describe the _____ static cosmology.

18. _____ _____ result when space is curved.

19. The _____ _____ assumes that the universe is the same everywhere at one time.

20. If we have a(n) _____ _____ , then the Pythagorean theorem will not be satisfied.

*21. _____ is used to measure nearby star distances and planet distances.

22. All that exists is the _____.

23. _____ is the study of the origin of the universe.

1. universe
2. steady state
3. Hubble's relation
4. variable
5. continuous creation
6. Einstein
7. Olbers' paradox
8. Euclidean
9. open universe
10. closed
11. homogeneous
12. cosmology
13. radar (or astronomical triangulation)
14. static
15. hierarchical
16. big bang theory
17. Newtonian-Euclidean
18. non-Euclidean geometries
19. cosmological principle
20. curved space
21. parallax
22. universe
23. cosmogony

Chapter 27
Cosmogony: A Twentieth-Century Version of the Creation

Answers to Problems in the Text

1. a.
2. The big bang theories require the existence of this radiation, which was predicted before it was observed. Thus, the big bang theories had made a prediction rather than just describing what was already known. The text indicates isotropy, but a slight anisotropy has recently been reported. It seems to be due to the Earth's motion in the galaxy and the galaxy's motion in the local cluster.
3. Globular clusters are Population I stars—old—so the heavy elements (Si, Fe, O, and so on) that compose the Earth are very rare in them.
4. Identify spectral lines as red-shifted known lines. The lines would be absorption lines caused by processes in the galaxies; they would not correspond to individual stars. Take the red shift as cosmological and, hence, Doppler.
5. Just as a volume of gas cools as it expands (for example, there are devices on the market that chill and frost drinking glasses by means of expanding CO_2 gas), so has the universe. The radiation has been red-shifted from its high temperature to its present 3 K.
6. Each student's answer will be different. If your class is small or you can break up into groups, this would make an excellent discussion topic. Ask for explanations of answers to the second question. Do these answers apply only to astronomy or to other scientific research as well?

Advanced Problems

7. a. Since $V = Hr$, then $r = V/H$, where r is the distance of a galaxy from us and V is its speed. Compare this with the general equation for the distance traveled by a body, $r = vt$, where v is the velocity, t is the time of travel, and r is the distance traveled. By comparison, $t = 1/H$ (*note:* $1/H$ has the dimension of time) and $1/H$ is then a measure of the age of the universe. Of course, the universe's expansion must slow down (because of gravity), so today's value of H gives only an upper time limit.

b. $H = 90$ km/s/megaparsec

$$\frac{1}{H} = \frac{1 \text{ megaparsec/s}}{90 \text{ km}} \text{ has dimensions of } \frac{\text{length} \times \text{time}}{\text{length}};$$

that is, $1/H$ has dimensions of time.

c. megaparsec $= 10^6$ parsecs $= 3 \times 10^{22}$ m

$$\frac{1}{H} = \frac{3 \times 10^{22} \text{ m} \times \text{s}}{9.0 \times 10^4 \text{ m}}$$

$$= 3.3 \times 10^{17} \text{ s}$$

$$= 3.3 \times 10^{17} \text{ s} \times \frac{1 \text{ year}}{\pi \times 10^7 \text{ s}}$$

$$= 1.1 \times 10^{10} \text{ years} = 11 \text{ billion years}$$

Sample Test Questions

True-False

1. Globular clusters are believed to be about 14 billion years old. T
2. If *H* is Hubble's constant, then *H* can also be called the expansion age. F
3. The universe is believed to be between 9 and 18 billion years old. T
4. Uranium production models show that the universe must be no older than 7 billion years. F
5. The universe began, according to big bang theories, as a low-density, high-temperature object. F
6. When the universe began, it was composed of hydrogen. F
7. The first atoms to condense from the fireball were helium. F
8. At the end of the first hour, the universe was about 75% hydrogen by mass. T
9. The formation of galaxies began in the universe a few billion years after the big bang. F
10. The heavy elements are believed to have been made mostly inside massive stars. T
11. At present, we believe the universe will expand forever. T

Multiple Choice

1. A myth is
    A. a falsehood
    *B. a scenario widely repeated and widely believed
    C. a star of mass 37 M
    D. both A and B
    E. all three, A, B, and C

2. The ages of globular clusters are _____ years.
    A. 4.6 billion   D. 1 million
    *B. 12 (±3) billion   E. 20 billion
    C. 16 (±4) billion

3. The elements in the solar system were incorporated into rocks _____ years ago.
    *A. 4.6 billion   D. 1 million
    B. 14 (±4) billion   E. 20 million
    C. 16 (±4) billion

4. The age of the universe based upon Hubble's constant is _____ years.
    A. 4.6 billion   D. 1 million
    B. 7 (±3) billion   E. 20 billion
    *C. 13 (±2) billion

5. From outside, the primeval fireball would have looked small, hot, and dense.
   A. correct
   B. wrong: small, hot, and rarefied
   C. wrong: large, hot, and dense
   D. wrong: large, cool, and dense
  *E. misleading: The fireball probably filled all space.

6. In the beginning, the universe was
   A. radiation   *D. both A and B
   B. subatomic particles   E. all three, A, B, and C
   C. atoms

7. The first atom to form as a fireball cooled was
  *A. hydrogen   D. uranium
   B. helium   E. librium
   C. deuterium

8. It is believed that at the end of the first hour of the universe the distribution was
   about _____% hydrogen, by mass, and _____% helium-4, by mass.
   A. 25/75   D. 90/10
  *B. 75/25   E. 10/90
   C. 50/50

9. If c is the speed of light and G is Newton's gravitational constant, then the energy
   equivalent of a mass M is
   A. Mc *D. $\overline{Mc^2}$
   B. M/c^2 E. $\sqrt{Mc/G}$
   C. GM/c^2

10. It is believed that the primeval radiation of millions of degrees has been red-shifted
    to _____ K.
    A. 273   D. 2
    B. 20   E. 0.01
   *C. 3

11. The heavy elements were first made
    A. in the primeval fireball
   *B. in the first massive stars
    C. when the density of the universe dropped enough for matter to form
    D. 0.001 second after creation
    E. misleading: never "first" made, always existed

*12. (Optional Equation IV) At what wavelength does the 3-K fireball radiation peak (approximately)?

A. 3 cm   D. 9.7×10^{-5} cm
B. 0.97 cm   E. 0.03 cm
*C. 0.097 cm

Essay

1. Describe at least two astronomical observations consistent with our estimated age of the universe. (See pp. 600–601.)
2. How are the heavy elements formed? (See p. 604.)
3. What existed before the big bang? (See p. 611.)
4. Do we know how, or if, the universe originated? Defend your answer.
5. Where did the carbon in our bodies originate? (See pp. 609–610.)
6. Justify the statement, "We are but stardust." (See pp. 609–610.)
7. What observations does the inflationary big bang model of the universe explain? (See pp. 602–603.)

Word Practice

1. The 3-K radiation is believed to be the red-shifted remnant of the

   _____ _____ .

2. The _____ _____ _____ _____ is placed at 10–18 billion years.

3. The _____ _____ _____ began a few million years after the big bang.

4. The temperature of the universe is about _____ _____ Kelvin.

5. A(n) _____ _____ big bang cosmology has the universe alternately expanding and contracting.

6. _____ _____ were created by nuclear reactions in stars.

7. The _____ _____ can be used to calculate the temperature of the big bang.

8. _____ is the study of the origin of the universe.

*9. _____ is the study of the structure of the universe.

10. The big bang theory allows us to calculate the _____ _____ of the universe.

11. The _____ of globular clusters in our galaxy are all placed at about 14 billion years.

12. The _____ of the universe may continue, or the universe may recollapse in the future.

13. _____ _____ were formed from stars.

14. The _____ of the galaxies could have occurred when the temperature and pressure of the universe had dropped to where gravitational contraction could occur.

15. The _____ _____ of the universe is the reciprocal of Hubble's constant.

16. The _____ of the elements were affected by supernova release.

17. _____ constant gives the expansion age of the universe.

18. A period of rapid _____ near the beginning of the big bang can explain many puzzling properties of the universe.

19. An early, extremely rapid expansion of the universe is proposed in the _____ big bang theory.

Answers to Word Practice

1. primeval fireball
2. age of the universe
3. formation of galaxies
4. 3 degrees
5. oscillating universe
6. heavy elements
7. 3-K radiation
8. cosmogony
9. cosmology
10. initial conditions

11. ages
12. expansion
13. heavy elements
14. formation
15. expansion age
16. abundances
17. Hubble's
18. inflation
19. inflationary

Chapter 28
Life in the Universe

Answers to Problems in the Text

1. Students need to look up the data for these stars. For example, what spectral type and population are they?

   Alpha Centauri is a binary system with G2 and K0 main-sequence stars of absolute magnitudes 4.4 and 5.7, respectively. γ Cephei is K1 subgiant of absolute magnitude 2.2. Sirius is another binary system consisting of an A1 main-sequence star and a white dwarf of absolute magnitudes 1.4 and 11.6, respectively.

   Of these stars, two (Alpha Centauri and Sirius) are binary stars, which reduces the chances of stable planetary orbits in the "habitable" zone. The companion of Sirius is a white dwarf that once must have been a giant, so if there is a planet upon which life existed, it probably would have been wiped out during that giant phase. Sirius itself is a large A1 star, so a near orbit to Sirius would be too hot (probably). There is a better chance for the Centauri system since both stars are relatively small and on the main sequence.

   γ Cephei is a single star and according to Table 28-1 may have a planetary companion of Jupiter's mass. However, γ Cephei is a subgiant probably in the process of evolving off the main sequence, so its "habitable" zone is changing. Life on a planet about γ Cephei would be in environmental trouble. In order of increasing probability of detecting signals, the choices are Alpha Centauri, Sirius, and γ Cephei, based solely upon distance. There is a good argument that Alpha Centauri, being a G2 star, has a greater probability of life than γ Cephei. The question is: Are Alpha A and Alpha Centauri B separated enough to allow good life-supporting orbits between?

2. Climatic changes have resulted from plate tectonics. "Outgassing" has altered atmospheric composition and has released some of our water supply. Some biologists attribute the extinction of some species to sudden climate changes, perhaps associated with continental drift and changing ocean currents.

3. As a star evolves, its total light output changes. Thus, over eons, the surface temperature will change dramatically. Even small solar changes may radically affect climates and life. As the star ages, it expands, heating the planet; later it cools, freezing the planet. If the planet's star explodes, all life is lost. Also, nearby stars might explode, bathing the planet in intense radiation that could affect existing life. Tidal forces can change spin rates. Orbital variations caused by perturbations change climate and may start ice ages.

4. Scenarios will vary—some possible examples:
   a. 1) International tensions build; nuclear war wipes out all life.
      2) Technology develops new bacterial weapons that get out of hand.
      3) Widespread use of new medicine. Long-term effects are genetic damage, killing all people.
      4) Industrial mistakes, such as widespread use of freon, alter climate suddenly.

   b. Space travel developed. Materials and energy obtained from other planetary systems, enabling the human species to survive critical damage to the Earth.

5. Answers will vary from student to student. At least two possible answers are given in each case. Of course, student answers should be much more detailed.
   a. 1) Panic; attempt made to destroy the aliens; war results.
      2) We go the way of the American Indians.
      3) They become our saviors; food and material problems solved; millennium arrives.
   b. 1) Communication established; information exchange helps both cultures.
      2) Communication established; information causes one civilization to collapse.
      3) Only one nation controls communication on this planet; war starts over fear of unilateral control of the incoming information.
   c. 1) Space colonization as we expand.
      2) Space travel forgotten; people turn inward.
      3) Other possibilities (justify).
6. Summary of some key points:
   Political situation unstable, especially as fossil fuels and limited resources are used up. Technology has potential of both destruction and space travel.
   One or the other (or both, successively) should occur; might conclude that since we are a relatively young planet, this spread has already occurred in the galaxy.
7. Course 2 is safest.
   a. If others chose Course 1, we would hear a lot. If they also concluded 2 or 3 was best, we would not hear much; everybody would be listening to nothing.
   b. No. If all adopted Course 2 or 3, nothing would be heard. If we believe travel is ultimately impossible, then Course 1 is safe. After all, those in charge could keep secret what they learned (if social stability is desired).
8. a. Too short-lived; no time for life to evolve.
   b. Evolution of star has altered region in which a planet could be suitable for life. Star is much brighter than originally.
   c. Star much dimmer than originally—all initial life-supporting regions now frozen. Moreover, the star would have already gone through a hot giant stage.
   d. Wow! Supernova occurred very nearby in past—that's it.
9. Mercury: No atmosphere or free liquid medium (H_2O) radiation.
   Venus: No liquid medium; temperatures too high for most organic processes.
   Earth: Intelligent life may yet evolve.
   Mars: Water frozen now; no long-lasting liquid medium; temperature extremes; apparently no abundant or sustained supply of liquid H_2O.
   Jupiter: Cold; updrafts from dense lower regions to top of atmosphere.
   Saturn, Uranus, Neptune, and Pluto: Too cold for biological reactions.
10. a. Not much change for 5–6 billion years, then will change rapidly in a few million years.
    b. Long.
    c. Possibly yes. Social structure might preserve existing biological structure (this eventuality is unlikely).
    More probably, no. Gradual, slow biological evolutionary changes will accumulate in at least 1 million years. Social and biological structures are transient.
    d. Might leave solar system and venture out in interplanetary or interstellar craft. Might have spread to many permanent colonies in distant space or in other planetary systems.

11. Planet diameter = 1.2×10^7 m, distance = 3×10^{16} m.

    a. $\alpha'' = (d/D) \times 2.06 \times 10^5 = \dfrac{1.2 \times 2.06}{3} \times 10^{7-16+5}$

      $\alpha'' = 8.2 \times 10^{-5}$ seconds of arc

    b. No.

    c. $\alpha'' = \dfrac{1.5 \times 2.06}{3} \times 10^{11-16+5}$

      $\alpha'' = 1''$

    d. Yes, in principle, but not in practice because the star will be so much brighter than the planet that to show the planet we would have to so overexpose the star that the effective resolution would be much worse than 1". (These latter comments may be missed by most students.)

12. a. Distance = 8200 parsecs

      Round trip   = 16,400 parsecs

                = 16,400 parsecs $\times \dfrac{3.26}{\text{parsec}}$ light-years

                = 5.35×10^4 light-years

      Travel time   = 53,500 years

    b. M13 is a Population I cluster: too old for much in the way of heavy elements. Supermassive stars that quickly evolved may have expelled heavy elements, but judging from the spectra of existing stars, no such elements are used in second-generation stars (or planets).

13. Diameter = 84 AU   = $8.4 \times 10 \times 1.5 \times 10^{11}$ m

                   = 1.3×10^{13} m

    Speed = $0.99 \times 3 \times 10^8$ m/s $\sim 3 \times 10^8$ m/s

    Time = $\dfrac{1.3 \times 10^{13}}{3 \times 10^8} = 4.3 \times 10^4$ s

        = $\dfrac{4.3 \times 10^4}{60 \times 60}$ hours = 12 hours

    a. In these short 12 hours, the Doppler shift would go from extreme blue shift to extreme red shift.

      We must use the relativistic Doppler shift formula (Optimal Basic Equation IX on p. 572) with $v/c = .99$. That is,

$$t = \Delta\lambda/\lambda = \sqrt{\dfrac{1 + v/c}{1 - v/c}} - 1.$$

When approaching, $v/c = -.99$;
while leaving, $v/c = +.99$.
Hence, $z = -.93$, $\lambda_{obs} = 1.5$ cm when the spacecraft is approaching; and $z = 13$, $\lambda_{obs} = 294$ cm when the spacecraft is receding. ($\Delta\lambda = \lambda_{obs} - \lambda_{true} = Z \; \lambda_{true}$)

    b. By the time any one frequency contact was recognized and the path established, the craft would be gone.

An even greater problem exists: Suppose the spaceship began transmission when just passing Pluto. The spaceship would take $14/84 \times 12 \sim 6$ hours to pass the Earth's orbit, while the light signal would take only 1% less time. That is, the signal would arrive about 3.5 minutes before the spacecraft itself. The actual detection is not more difficult, but it happens so rapidly that we might miss it.

Sample Test Questions

True-False

1. Life is a process. T
2. Most of your body weight is water. T
3. Proteins are composed of amino acids. T
4. The surface of the Earth is screened from ultraviolet rays by the ozone layer. T
5. The greater production of offspring by those individuals best adapted to the changing environment is called natural selection. T
6. The statistics of masses among binary and multiple stars suggest that few stars are likely to have companions. F
7. Massive stars spin more slowly than smaller stars. F
8. For life to evolve on a planet, the planet must be close enough to its star that water is a liquid but not so close that the water all evaporates. T
9. No Earth organisms could survive under any conditions found on other planets. F
10. Contact with an extraterrestrial civilization would pose no danger to humanity. F
11. Extraterrestrial intelligent life definitely exists in the solar system. F
12. There is the possibility of primitive life forms elsewhere in the solar system. T
13. Life, even intelligent life, is very probable elsewhere in the universe. T
14. Evolution may pass through only a brief explorative interval in which societies on one planet would care to reach other planets. T
15. Good evidence exists for the "ancient astronaut" hypothesis. F
16. The first radio message deliberately sent to the stars was sent toward M13 in 1974. T

Multiple Choice

1. The chemistry of carbon compounds is termed
   A. cell chemistry
   *B. organic chemistry
   C. biochemistry
   D. a minor chemistry
   E. none of these

2. Which of the following suggest(s) water was crucial to the development of life on Earth?
   A. People, animals, and plants are mostly water.
   B. Organisms without water quickly die.
   C. Iron is rusted by water.
   *D. Both A and B suggest water's importance.
   E. All three, A, B, and C, suggest its importance.

213

3. One of the first scientists to produce amino acids using electricity and a mixture that might approximate the primitive atmosphere was
   A. Oparin
   B. Fox
   C. Drake
   D. Shklovskii
   *E. Miller

4. H. G. Bungenberg de Jong found a spontaneous synthesized object of about 1–100 microns, rich in colloidal matter and clearly separated from the external environment. Such objects are termed
   A. proteinoids
   B. microspheres
   *C. coacervates
   D. colloidal drops

5. S. W. Fox has shown that if dry amino acids are heated and water is added, cell-like objects called _____ are formed.
   *A. proteinoids
   B. microspheres
   C. coacervates
   D. colloidal drops

6. Which of the following is (are) evidence for planets near other stars?
   *A. Massive stars spin faster than smaller stars.
   B. Sirius is a binary star.
   C. The Sun is not special, so other stars have planets.
   D. Both A and C are evidence.
   E. All three, A, B, and C, are evidence.

7. It is unlikely that the oldest stars (Population II) have planets because
   A. they are too bright
   B. they are too big
   C. they contain too much of the heavy elements
   *D. they contain too little of the heavy elements
   E. they are too red

8. Binary stars are less likely to have habitable planets because
   A. most have widely differing masses
   B. the Lagrangian surface lies between them
   *C. many possible orbits suffer large perturbations, causing orbits to be highly elliptical if not parabolic
   D. both B and C
   E. all three, A, B, and C

9. Biologists concerned with possible life on other worlds are called
   A. crazy
   B. extraterrestrial biologists
   *C. exobiologists
   D. technocrats
   E. ecdysiasts

10. There is probably _____ intelligent life in the solar system, according to the text.
   A. no
   *B. no extraterrestrial
   C. permanent extraterrestrial
   D. visiting interstellar

11. The estimated fraction of stars with planets having intelligent life ranges from a low of _____ to a high of _____.
   A. $10^{-20}/10^{-14}$
   *B. $10^{-14}/2 \times 10^{-2}$
   C. $2 \times 10^{-2}/3 \times 10^{-1}$
   D. $.03/.1$
   E. $1/30$

12. If we accept the plausible upper limit of the estimated fraction of stars with planets having intelligent life and take the number of stars in the galaxy as 100 billion, we get _____ worlds with intelligent life.
   A. 2
   B. 2000
   C. 2 million
   *D. 2 billion
   E. 2000 billion

13. If we accept the text's plausible upper limit on the fraction of stars with planets having intelligent life, we get the distance to the nearest civilization as _____ light-years.
   *A. 15
   B. 30
   C. 150
   D. 1500
   E. 10^7

14. The spacecraft _____ flew by Jupiter in 1973 and will leave the solar system. It contains a plaque attempting to convey our appearance and location.
   A. Apollo 13
   B. Mariner 10
   C. Viking 2
   D. Pioneer 3
   *E. Pioneer 10

Essay

1. What is the evidence for planets near other stars? (See p. 621.)
2. What astronomical or planetary evolutionary processes could have affected biological evolution? (See pp. 623–625.)
3. What are four conditions needed to make a planet habitable? (Six are listed on p. 623.)
4. What evidence, and what statistical possibility, is there for alien life in the solar system? Intelligent life? (See p. 629.)

Word Practice

1. _____ are biologists concerned with possible life on other worlds.

*2. The _____ nature of life gives us clues to the processes involved in its origins.

3. The molecules formed of amino acids in cells are _____.

4. _____ _____ are any complex carbon-based molecules.

5. The _____ layer screens the Earth's surface from ultraviolet radiation.

6. It will be necessary to surmount a(n)_____ _____ if intelligent life is to survive on Earth.

*7. All life on Earth is _____-based.

*8. Life is a(n) _____ rather than a state.

9. _____ _____ are the building blocks of proteins.

10. Evolution may pass through a brief _____ _____ when communication between societies on different planets would be desired.

11. _____ is a process, not a state.

*12. _____ is probably the best method of contact with alien civilizations.

13. The _____ _____ protects the Earth's surface from excessive ultraviolet radiation.

14. _____ are cell-like clusters of organic molecules discovered by H. G. Bungenberg de Jong.

15. _____ _____ is the chemistry of carbon molecules.

16. The _____ is the chemical factory upon which life is based.

17. The _____ _____ produced amino acids under conditions believed to approximate the early Earth.

18. _____ are round, cell-like objects created by S. W. Fox that grow and divide but are not considered living.

19. There are four kinds of _____ for planets near other stars.

20. _____ selection affects evolution because those individuals best adapted to the changing environment are more likely to have offspring.

21. Natural _____ eliminates those species not suited to their environment.

*22. _____ _____ was the first spacecraft launched that should leave the solar system.

*23. It is almost certain that there is no intelligent _____ life in our solar system.

24. The _____ _____ enables us to calculate the probable distance to another intelligent species.

25. The fossils called _____ are the oldest compelling evidence for life on Earth.

Answers to Word Practice

1. exobiologists
2. dynamic
3. proteins
4. organic molecules
5. ozone
6. cultural hurdle
7. carbon
8. process
9. amino acids
10. explorative interval
11. life
12. radio
13. ozone layer
14. coacervates
15. organic chemistry
16. cell
17. Miller experiment
18. proteinoids
19. evidence
20. natural
21. selection
22. Pioneer 10
23. alien
24. Drake equation
25. stromatolites

The Cosmic Perspective

Test Suggestions

Give the students any one of the 12 points covered—a brief phrase similar to those following. Ask them to elaborate, either agreeing or disagreeing, and to support their viewpoint.

1. Astronomy is important.
2. The Earth is a spaceship.
3. We must take care of our Earth, whether it is unique or not.
4. We are not living within our planetary means.
5. We can engineer a stable society.
6. We understand the structure of our universe.
7. We can know what is happening in the universe.
8. The universe can exist without us.
9. Nothing is permanent. (What does *permanent* mean?)
10. Humanity has changed drastically in the past and will in the future.
11. Much of the universe is beyond our power to control.
12. There is not much left for science to understand, and what we do not understand is not worth the effort.

Essay A
Pseudoscience and Nonscience

It is interesting to quiz students on this material at the start of the course and compare their answers with their responses after studying the material.

Sample Test Questions

True-False

1. A pseudoscience has a consistent body of supporting evidence. F
2. The processes of perception, conception, and reporting always produce correct data. F
3. UFO literature must be pseudoscientific. F
4. The so-called 1908 Siberian meteorite impact was really a nuclear explosion. F
5. Velikovsky is clearly correct in saying that heavenly catastrophes occurred in 1500 B.C. and 750 B.C. F
6. "Scientific creationism" is a science. F

Multiple Choice

1. A _____ is a body of hypotheses treated as true but without any consistent body of supporting experimental or observational evidence.
   - A. science
   - *B. pseudoscience
   - C. happening
   - D. horoscope
   - E. none of these

2. Pseudoscience is dangerous because
   - A. consumers often lose their money on false promises
   - B. real scientific discoveries are misrepresented
   - C. it contributes to anti-intellectual attitudes that exchange mysticism and magic for explanation and discoveries
   - D. both A and C
   - *E. all three, A, B, and C

3. The steps needed for an eyewitness report include
   - A. perception
   - B. reporting
   - C. interpretation
   - *D. both A and B
   - E. all three, A, B, and C

4. The second step in generating an eyewitness report is
   - A. reporting
   - B. perception
   - *C. understanding
   - D. none of these

5. All UFO literature must be pseudoscientific.
   - A. correct
   - B. wrong: All is except that published by professional astronomers.
   - *C. wrong: The judgment must be based upon how evidence is treated.

219

6. The following characteristics distinguish scientific creationism from science.
    A. advancing definite knowledge of what is correct before observations are made
    B. the process of selecting only those facts that are desired
    C. the acceptability of the observed results to the scientific establishment
   *D. both A and B
    E. all three, A, B, and C

Essay

1. What evidence is there for ancient astronauts?
2. Why are Velikovsky's ideas termed "pseudoscience"?
3. Explain why "scientific creationism" is not a science.

Astronomical Coordinates and Timekeeping Systems

Sample Test Questions

True-False

1. Right ascension is measured from the vernal equinox. T
2. A catalog of right ascensions with epoch 1950 indicates that all values given are good until 1950. F
3. Atomic time is based on the cesium 133 atom. T
4. Intercalculation is the process of adding extra days to the calendar to keep it in step with astronomical events. T
5. A Julian day count is the count of days since January 1, 1776 A.D. F

Multiple Choice

1. The sidereal day is defined as the
   *A. interval between successive transits of the vernal equinox
   B. interval between successive sunrises
   C. hour angle of the Sun + 12 hours
   D. hour angle of the mean Sun + 12 hours
   E. none of the above

2. The zenith is the point
   A. where you are
   *B. in the sky directly overhead
   C. in the sky directly below you
   D. in the sky over Greenwich, England
   E. none of the above

3. The ecliptic is
   A. the line midway between the north and south celestial poles
   *B. the Sun's apparent path against the background stars
   C. the Moon's orbit as seen against the background stars
   D. an event when the Moon passes in front of the Sun
   E. none of the above

4. The equator system uses the celestial equator as the fundamental plane and the vernal equinox as the reference point; its coordinates are right ascension and declination. It is the most convenient astronomical coordinate system to use in following a star during a night.
   *A. correct
   B. wrong: The galactic system is the best for tracking.
   C. wrong: It is the horizon system that fits this description.
   D. wrong: It is the metric system that fits this description.
   E. none of the above

5. Declination is from what coordinate system?
   *A. equator
   B. horizon
   C. metric
   D. galactic
   E. economic

6. The coordinate right ascension is usually measured in
   A. degrees, minutes, seconds
   *B. hours, minutes, seconds
   C. radians
   D. meters
   E. AUs

7. At the vernal equinox, the Sun
   *A. crosses the equator from south to north
   B. crosses the equator from north to south
   C. is at its northernmost point
   D. is at its southernmost point
   E. has a right ascension of 12 hours

8. The reference direction in the horizon system is the
   A. vernal equinox
   B. north celestial pole
   *C. north point
   D. zenith
   E. none of these

9. The coordinate altitude is from the _____ system.
   A. galactic
   B. equator
   *C. horizon
   D. ecliptic
   E. metric

10. Apparent solar time is the
    A. interval between successive transits of the vernal equinox
    B. hour angle of the vernal equinox
    C. hour angle of the Sun
    *D. hour angle of the Sun + 12 hours

11. The sidereal year is based upon the fixed stars. It is the most convenient year to use because it keeps pace with the seasons.
    A. correct
    B. wrong: The anomalistic year is the most convenient because it keeps pace with the seasons.
    *C. wrong: The tropical year is the most convenient because it keeps pace with the seasons.
    D. wrong: The sidereal year is based upon perihelion passages.
    E. both C and D

222